The Bradley and How It Got That Way

Recent Titles in
Contributions in Military Studies

War for the Every Day: Generals, Knowledge, and Warfare in Early Modern Europe, 1680–1740
Erik A. Lund

Desert Shield at Sea: What the Navy Really Did
Marvin Pokrant

The Moroccan Goums: Tribal Warriors in a Modern War
Edward L. Bimberg

Military Assistance: An Operational Perspective
William H. Mott IV

Desert Storm at Sea: What the Navy Really Did
Marvin Pokrant

A Peripheral Weapon? The Production and Employment of British Tanks in the First World War
David J. Childs

Where a Man Can Go: Major General William Phillips, British Royal Artillery, 1731–1781
Robert P. Davis

Truman, MacArthur, and the Korean War
Dennis D. Wainstock

Secret Soldiers of the Revolution: Soviet Military Intelligence, 1918–1933
Raymond W. Leonard

The Writing of Official Military History
Robin Higham, editor

Reorganizing the Joint Chiefs of Staff: The Goldwater-Nichols Act of 1986
Gordon Nathaniel Lederman

Bullies and Cowards: The West Point Hazing Scandal, 1898–1901
Philip W. Leon

The Bradley and How It Got That Way

Technology, Institutions, and the Problem of Mechanized Infantry in the United States Army

W. Blair Haworth, Jr.

Contributions in Military Studies, Number 180

GREENWOOD PRESS
Westport, Connecticut • London

Library of Congress Cataloging-in-Publication Data

Haworth, W. Blair, 1962–
The Bradley and how it got that way : technology, institutions, and the problem of mechanized infantry in the United States Army / W. Blair Haworth, Jr.
p. cm.—(Contributions in military studies, ISSN 0883–6884 ; no. 180)
Includes bibliographical references and index.
ISBN 0–313–30974–4 (alk. paper)
1. Mechanization, Military—United States—History. 2. United States. Army—Organization. 3. United States. Army. Infantry. 4. M2 Bradley infantry fighting vehicle. I. Title. II. Series.
UG446.5.H374 1999
356′.16—dc21 99–18594

British Library Cataloguing in Publication Data is available.

Library of Congress Catalog Card Number: 99–18594
ISBN: 0–313–30974–4
ISSN: 0883–6884

First published in 1999

Greenwood Press, 88 Post Road West, Westport, CT 06881
An imprint of Greenwood Publishing Group, Inc.
www.greenwood.com

Printed in the United States of America

The paper used in this book complies with the Permanent Paper Standard issued by the National Information Standards Organization (Z39.48–1984).

10 9 8 7 6 5 4 3 2 1

To Jerry Haney, Stephen Baxter, and Theodore Ropp,
for getting me into this;
and to Maggie,
for getting me through it.

Contents

A photo essay follows page 93

Acknowledgments

This work has been about seven years in the making since its conception as a doctoral dissertation in the Department of History at Duke University; I have a lot of thanks stored up:

Professors Alex Roland and I. B. Holley, Jr. have been unfailing sources of insight and support for this project; Professors Lawrence Goodwyn, Tami Davis Biddle, and Albert F. Eldridge have provided valuable input as well. The staff of Perkins Library, Duke University have been indispensable, in particular those of the Reference, Public Documents & Maps, and Collection Development Departments. The latter has been my employer the whole time, and I must especially thank my department head, Ms. Virginia Gilbert, and my colleagues for their flexibility and forbearance. The staff and collections of Davis Library of my alma mater, the University of North Carolina at Chapel Hill, have been invaluable as well. The staff of the National Archives have been unfailingly helpful, especially Mr. Richard Boylan. So have the United States Army research facilities that I have availed myself of: the Center of Military History, the Military History Institute, and the historical offices of the Infantry Center, the Armor Center, and the Materiel, Training and Doctrine, and Tank-Automotive Commands. Command historian Dale Prentiss and Ms. Ann Bos of the latter were especially helpful in providing photographs, as was Mr. James Carter of United Defense L. P.

A variety of extra-academic communities have lent support for this project as well: the USENET sci.military.moderated newsgroup has provided a valuable source of raw information and informants. Furthermore, it has been a testing ground for many of the hypotheses and explications that have gone into this work: an audience of thousands, many of them well-informed and more of them vocal, all with the ability to reply to articles instantly, is a powerful sanity check. The USENET alt.peeves newsgroup has provided the latter service as well, albeit in a decidedly different fashion. He's Not Here, Chapel Hill, NC, must also be mentioned as providing a supportive environment for the conduct of clerical, editorial, and research chores associated with writing on those days when a four-million-volume library contains at least a million pressing distractions.

My friends have put up with years of preoccupation without complaint. My parents and my immediate and extended family have provided greater amounts of more kinds of help than can easily be listed. My wife, Maggie—a librarian by training, an editor by profession, and profoundly patient, thoughtful, and analytical by nature—has provided greater amounts of more kinds of help than can conceivably be listed. I am the most fortunate of men.

I write this the day after Memorial Day, 1999, which evokes a debt that can't be repaid. I hope this book is at least good for a little interest.

Abbreviations

ACAV	Armored Cavalry Assault Vehicle
AFV	Armored Family of Vehicles
AIFV	Armored Infantry Fighting Vehicle
AOE	Army of Excellence
APC	armored personnel carrier
APDS	Armor-Piercing Discarding Sabot
APFSDS	Armor-Piercing Fin-Stabilized Discarding Sabot
AR/AAV	Armored Reconnaissance/Airborne Assault Vehicle
ARSV	Armored Reconnaissance Scout Vehicle
ARVN	Army of the Republic of Vietnam
BAR	Browning Automatic Rifle
BFV	Bradley Fighting Vehicle
BRL	Ballistic Research Laboratory
CAMB	Combined Arms Maneuver Battalion
CDC	Combat Development Command
CFV	Cavalry Fighting Vehicle
CONARC	Continental Army Command
DRS	Divisional Restructuring Study
FIFV	Future Infantry Fighting Vehicle
FIV	Future Infantry Vehicle
FIST	Fire Support Team
FPW	Firing Port Weapon
FVSC	Fighting Vehicle Systems Carrier
FY	Fiscal Year
GAO	General Accounting Office
HE	high explosive
HFM	Heavy Forces Modernization

HHC	Headquarters and Headquarters Company
HMMWV	High Mobility Multipurpose Wheeled Vehicle
IDF	Israeli Defense Forces
IFV	Infantry Fighting Vehicle
ILS	Integrated Logistic System
IRUS	Infantry Rifle Unit Study
ISU	Integrated Sight Unit
IVIS	Intervehicular Information System
LID	Light Infantry Division
MBT	Main Battle Tank
MICV	Mechanized Infantry Combat Vehicle
MN	Material Need
MOS	Military Occupational Specialty
MTOE	Modified Table of Organization and Equipment
MTW	(German) *Mannschaftstransportwagen,* armored personnel carrier
NATO	North Atlantic Treaty Organization
NBC	Nuclear Biological and Chemical Warfare
NET	New Equipment Training
OTAC	Ordnance Tank-Automotive Command
QMR	Quantified Material Requirement
RFP	Request for Proposal
ROAD	Reorganization Objective Army Division
SSG	Staff Sergeant
TACOM	Tank-Automotive Command
TBAT-II	TOW-Bushmaster Armored Turret, Two-man
TRADOC	Training and Doctrine Command
U-COFT	Unit Conduct of Fire Trainer
VC	Viet Cong
VRFWS(-S)	Vehicle Rapid-Fire Weapons System (-Successor)
VT	"variable time"

Introduction

> D Company, TF4-7th Infantry, had moved out that morning after a brief pause to rest and refuel. The company reached Medina Ridge after only a 5-kilometer march, and Staff Sergeant Charles Peters, the company master gunner, spotted several BMPs and T-72s dug in on the slope below his Bradley. Maintaining the strictest fire control, Peters destroyed one BMP with only three quick rounds of 25mm armor-piercing ammunition, then adroitly switched his ammunition selector to high explosive to engage the Iraqis as they ran from the vehicle toward some nearby trenches. He just as rapidly switched back to armor-piercing again to kill a second BMP and engaged a third, raising a cloud of dust all around the target with impacting 25mm cannon rounds. This particular Iraqi vehicle strangely did not explode like the others, and as the dust settled, Peters saw why—it was a T-72. As the tank opened fire on his company, Peters quickly raised the TOW "two-pack" launcher and switched sighting systems a third time. The sequence took only seconds, but it seemed like hours before he could launch the missile. Peters held his cross hairs steady on the tank a few seconds longer until missile impact and destruction. He finished this remarkable one-man gunnery demonstration by switching back to the 25mm a fourth time to kill a third BMP as Captain Christopher Shalosky, his amazed company commander, watched.
>
> —Robert H. Scales, Jr.[1]

This official account from the third day of the ground phase of Operation DESERT STORM describes a remarkable moment in armored warfare, to be sure. It is the more remarkable when the reader considers that Staff Sergeant Peters, an infantry noncommissioned officer in an infantry unit tracing its lineage back to 1812, was nonetheless fighting from an armored vehicle. For infantrymen to fight—and moreover, to train to do so as a matter of course—in this manner seems counterintuitive, to say the least. Their employment in the role of mechanized infantry is the result of a complex set of human, organizational, and technological processes that merits further study.

One avenue for such further study presents itself in the instrument SSG Peters used to accomplish his feat of arms, the Bradley infantry fighting vehicle (IFV). It took the United States Army twenty years to get an IFV into the field; once there, it was the source of considerable internal and external controversy. What made the Bradley Fighting Vehicle so controversial, and its development so difficult? Informed discussion of the problems of the Bradley are inextricable from discussion of the larger idea of mechanized infantry in the U.S. Army. It is a complex pair of subjects; as will be seen, even the term "mechanized infantry"—used by default for infantry forces equipped with specialized armored transport—is freighted with hidden meanings and agendas, and is, in fact, a relatively recent coinage. A discussion of the subject requires extensive description of the development of the equipment and organizations involved, as well as their extra-American counterparts. These tend to be both complicated and unfamiliar to the typical reader, yet they are vitally important to the issue involved, which is fundamentally the institutional affinities of mechanized infantrymen. Simple *motorization* of infantry, the integration of relatively standard motor vehicles into infantry units, is fairly straightforward; motorized infantry units fight in similar fashion to their lighter-equipped counterparts, although their logistics change considerably. With *mechanization*, as infantry vehicles become more capable, more specialized, and more intimately involved with tank forces, new questions arise. Is the mechanized infantry force a body of infantrymen who happen to be issued armored vehicles, or are they armored vehicle crewmen who happen to dismount for some combat situations? Conversely, is the main weapon of the vehicle the troops carried within, or the armament mounted on board? These are the most important subsidiary questions in any inquiry into the Bradley Fighting Vehicle's turbulent history: doctrine and tradition has said mechanized infantrymen are the former; practice and the thrust of technological development has tended to make them the latter, and the effort to reconcile the two divergent views in the same soldiery has been a continual source of problems.

This dichotomy makes for a rather frustrating historical exercise. As may be seen from the bibliographic essay following the main body of this work, many traditional avenues of investigation are complicated by the question of institutional affiliation: the student of the field is therefore often on difficult theoretical and historiographic ground, especially given the archival and security problems accompanying any project in recent military history. Useful sources of information may nonetheless be found, and combining a study of these sources with more traditional ones, it is possible to reach a thesis going beyond the mechanics of the U.S. military procurement process, which has largely dominated discussion of the problems of the Bradley Fighting Vehicle while obscuring those of the mechanized infantrymen equipped with it. The well-studied vagaries of the U.S. government's defense procurement process surely aggravated the problems of the Bradley and its precursors, but did not cause them. The U.S. Army had difficulty acquiring and accommodating an infantry fighting vehicle because it insisted on and persisted in effecting a radical doctrinal change by incremental means. Since the early days of the Armored Force just prior to World War II, the Army had maintained armored infantry units in its armored divisions, specialized to fight alongside its tank forces. As a matter of doctrine, these units had been trained and equipped solely for dismounted combat since the end of World War II. Their vehicles, generically referred to as armored personnel carriers (APCs), while protected by modest amounts of armor, were intended solely as a means of transport. By the end of the 1950s, it appeared that technological conditions—in particular, the threat of nuclear, biological, and chemical warfare—made this arrangement appear less tenable.

It was apparent that the new style of combat envisioned would therefore call for a new type of infantry vehicle, one capable of performing as a combat vehicle rather than as a relatively simple means of transport. The Army development apparatus, however, chose to derive the characteristics of the new infantry fighting vehicle—sometimes referred to as mechanized infantry combat vehicles (MICVs) and variations of these two terms—from those of the earlier armored personnel carrier. These characteristics included the demanding requirements to maintain a weight and size commensurate with airportability and amphibious capability while carrying an infantry squad and its equipment. Unfortunately, these were in direct conflict with the increased firepower, protection, and mobility needed to operate in conjunction with tanks. This demand for enhanced performance without compensatory trade-offs, which can evocatively be labeled "capability greed," led to a series of unacceptable designs over the course of the 1960s.

Capability greed was accompanied and exacerbated by a related imperative which may be called "standardization greed." The idea of "armored infantry," a relatively specialized force oriented to the tank, gave way to that of "mechanized infantry," a more generalized infantry force that used armored vehicles in the course of accomplishing the spectrum of infantry missions. This monizing tendency led to standardization of training and doctrine, and with it standardization of organization and equipment among the various sorts of infantry units. Standardization had long been regarded as a vital element of both training and procurement in the Army; paradoxically in this case, though, the demand for simpler organization overall led to more complex equipment, as requirements broadened to accommodate more missions within a single device. This phenomenon contributed further to the problems of procuring a new infantry vehicle.

As this situation developed, a similar problem was faced by armored cavalry units in the Army. The attempt to accommodate both the reconnaissance and combat missions of armored cavalry units through advanced technology, while at the same time reducing the number of types of fighting vehicles in armored cavalry units, had led to a similar string of unsuccessful vehicles. In the face of constrained development resources in the wake of the Vietnam War, the two troubled programs, cavalry and infantry vehicle, were combined into one. To accomplish this merger, many of the requirements complicating the earlier development efforts were compromised; none were abandoned. The resulting machine, by now named the Bradley Fighting Vehicle, finally reached service in 1983.

The new vehicle was generally popular with the infantry, who welcomed the enhanced capabilities it promised. At the same time, the branch was concerned by the training impact of such a complex piece of equipment and the problems encountered by Bradley-equipped mechanized infantry in accomplishing traditional infantry missions. The cavalry branch, on the other hand, found itself committed to carrying out both its reconnaissance and its combat missions with a large, heavily-armed, though not tanklike, vehicle. This environment of complexity, compromise, and controversy drew the unfavorable attention of numerous outside observers, many of whom focused on the vehicle's vulnerability relative to that of the tanks with which it was to work.

By the time of Operation DESERT STORM, the bulk of these problems had been addressed. In most cases, the Bradley's perceived problems were dealt with by abandoning the secondary capabilities and wide-ranging missions that had so complicated the development of the Bradley to begin with. It was, to a great extent, a matter of the Army as an institution shrugging off—or at least of its rationalizing—the problems of standardization greed and capability greed. Doing so within the framework of the service's traditional branches, their missions, and their culture cost the U.S. Army greatly in time, money, political capital, and, ultimately, in the effectiveness of the system it fielded.

NOTE

1. Robert H. Scales, Jr., *Certain Victory* (Washington, DC: Office of the Chief of Staff, United States Army, 1993), 296.

Chapter 1

Infantry and Tanks: Early Armored Warfare and the Origins of the Mechanized Infantry Idea, 1916–1939

> It is worth mentioning that the Victoria Crosses and other high decorations in which the [Tank] Corps prides itself were for the most part won by members of crews driven out of broken tanks fighting with dismounted Lewis guns in places that no infantryman could have reached.
>
> —A. J. Smithers[1]

To understand the problems involved in discussing the mechanized infantry idea, it is necessary to review briefly the history of armored warfare. The emphasis here differs from most accounts, which center on the tank. There is little need to discuss the prehistory involved, save to note that since the introduction of gunpowder weapons into the Western military tradition, concerted efforts have been made to produce a mounted infantry arm, combining the mobility of cavalry with the staying power of infantry. Historically, these troops have tended to gravitate toward employment as cavalry, as the demands and potentialities of mounted service asserted themselves. A notable exception to this has been the U.S. Army, which for much of its history—indeed, until the obsolescence of horse cavalry—tended to operate its nominal cavalry arm as mounted infantry.

It is tempting but misleading to trace the mechanized infantry phenomenon back this far. However, as the main armored warfare theorist to study the subject, Richard Simpkin, points out,[2] mounted infantry units per se left few institutional traces, generally being ad hoc conversions of infantry or cavalry units, each of which had their own reasons for regarding such employment as a sideshow. Furthermore, mounted infantry was fundamentally tied to the idiosyncrasies of the horse in its military characteristics, particularly the animal's vulnerability on the modern battlefield. The capacity for protected mobility—the ability to cross fire-swept ground in pursuit of tactical decision—changed the terms of mounted operations so much as to render comparisons misleading. It was the superimposition of the tank onto the environment of Western warfare that made the question of tactical mobility for infantry an urgent and perennial one; it is here that our story begins.

The tank began as an infantry support weapon. Available power plants and running gear, providing 105 horsepower to a 28-ton machine, kept the earliest tanks from moving much faster than men could walk;[3] rather, they moved at that speed when their overstrained, underdeveloped power plants and running gear worked well enough to move them at all. In their first engagement during the battle of the Somme (Bapaume, 15 September 1916), only eleven of the forty-seven machines brought up got into battle.[4] Their weapons, too, were chosen to engage field fortifications impeding the infantry's advance. Given these factors, tank-infantry cooperation was both relatively simple and absolutely essential to the successful prosecution of operations involving the new weapon. The rewards of success and the penalties of failure were both made readily apparent at the battle of Cambrai in November 1917.

Although tanks had been in service for over a year, Cambrai marked the first attempt to exploit the new weapon's capabilities as an integral part of the battle plan, rather than as an afterthought. At dawn on 20 November 1917, some 200 British tanks advanced in a sector chosen for favorable terrain—a firm, level area later compared by tank pioneer Percy Hobart to the interwar British armored maneuver area at Salisbury Plain.[5] Preliminary bombardment was kept to a minimum and relied on new "silent registration" techniques. This served both to achieve surprise and to preserve mobility by avoiding the boggy moonscape created by heavy artillery in previous battles.[6] Along most of a six-mile front, infantrymen practicing a specially devised battle drill advanced in file behind each section of three tanks, gaining some shelter from direct fire behind their armored bulk while protecting the rear of the lumbering, near-blind machines. Meanwhile, the tanks' Lewis machine guns and six-pounder cannon suppressed the German positions—destroying them, or at least preventing them from firing effectively—as their treads flattened barbed wire entanglements. The German defense was dislodged by this onslaught, and fell back over five miles.[7]

In the sixth mile, at Flesquières Ridge, the situation was different. There, the commander of 51 (Highland) Division, Major-General G. M. ("Uncle") Harper, disdained the new drill, sending his infantry in extended line some one hundred yards behind the tanks. Prevented from effectively cooperating with the tanks by this seperation, the infantry advance was soon bogged down by enemy fire, while the tanks blundered forward across the German trench lines only to be engaged by direct-firing field guns. These quick-firing weapons with their high explosive shells far outmatched the tanks' protection, intended as it was only to keep out shell fragments, shrapnel balls, and small-arms fire. Without infantry available to clear out the gun positions, twenty-seven tanks—over a company's worth—were knocked out.[8]

The failure at Flesquières combined with attrition among men and tanks elsewhere to produce an indifferent result; nonetheless, the lessons of Cambrai were not lost on the Tank Corps. Close cooperation with infantry was evidently vital to successful armored penetration of enemy lines, and protection of the infantry was just as evidently vital to close cooperation. How best, then, to provide the protection of the tank to infantrymen? The immediate answer was to put the infantry into the tank itself. The Tank Corps had had some involuntary experience with this method, since breakdowns or battle damage could easily turn a tank's eight-man crew into a de facto infantry section, with the results alluded to by A. J. Smithers:

> It is worth mentioning that the Victoria Crosses and other high decorations in which the [Tank] Corps prides itself were for the most part won by members of crews driven out of broken tanks fighting with dismounted Lewis guns in places that no infantryman could have reached.[9]

The cramped quarters of the first tanks made the idea of carrying infantry difficult to implement, but a serendipitous solution presented itself. The tank had been designed to cross trenches; one of the first countermeasures to the new threat was naturally enough the digging of wider trenches by the Germans. The quickest British response, equally naturally, was the fielding of a longer tank. The Mark V tank, intended for the spring 1918 campaigns, was lengthened to a total of over 32 feet to produce the Mk V*, the "star" representing a development of the basic model in British nomenclature. The addition of 6 feet and 5 tons to the vehicle made it unwieldy and underpowered even by the standards of the time (a Mark V** with a 225hp engine was developed to overcome this, but never fielded), but it did improve trench-crossing capability. It also encompassed another feature: space became available for twenty to twenty-five troops. Organized into two Lewis gun sections, they could be dismounted to help the tanks forward or to hold ground already taken.[10] This new tactical concept was tried at Amiens in August 1918. It failed.

The failure arose not because of any inherent unsoundness of the idea, but because the specialist planners of the Tank Corps failed to take into account the effect of the early tanks' environment on unacclimated troops. Conditions were atrocious: crew and passengers were installed in an unsprung, almost lightless and airless armored box where temperatures commonly reached 125 degrees Fahrenheit; there they were bathed in the fumes and noise of the unshielded, unmuffled engine and outgoing gunfire from the inside, and those of gas attacks and incoming gunfire from the outside.[11] Immersed in this environment on the approach, the embarked infantrymen were so overcome by heat, fumes, and motion sickness that they were largely unfit for action when the time came to dismount. Their failure was all the more pronounced given the relative success of Cambrai-style cooperation between tanks and infantry advancing under their own power in the same battle.[12]

Despite this state of affairs, the requirement for protected mobility for infantry working with tanks—that is, the ability to cross fire-swept ground—remained. Having determined the vulnerability of unsupported tanks to properly trained and equipped artillery and infantry, the Germans made separation of tanks from infantry and the defeat of each element in detail the foundation of their antitank tactics. These problems had to be dealt with: the Allied offensive being planned for 1919 would rely heavily on the use of tanks to break through the now-depleted German defenses. A new solution presented itself in the guise of the Mark IX, a specialized armored carrier vehicle derived from the Mark V tank. Stripped of armament and (mercifully) fitted with improved ventilation, the Mark IX could carry fifty infantrymen or ten tons of supplies, thus addressing another bottleneck in the conduct of nascent armored warfare, tactical logistics. Only thirty-five of these machines had been delivered by the Armistice. None saw action.[13]

By the Armistice, the British had encountered all the basic problems of armored warfare. Through the efforts of Tank Corps commanders such as Major-General Hugh Elles and staff officers such as then Lieutenant-Colonel J. F. C. Fuller, they had come to grips with them to a sufficient degree that the conduct of operational-level missions—campaigns featuring armored attacks into enemy rear areas—were envisioned. Similarly, the French had achieved some expertise, though they proceeded from different doctrinal assumptions and lacked either an Elles to direct their effort or a Fuller to plan it. The United States Army, meanwhile, was at nowhere near so high a conceptual level. As with much of the rest of the arms industry, the nation's late entry into the war and difficulties in getting production onto a war footing meant that U.S. forces operated only a handful of British Mark V and Mark VIII tanks, along with some French Renault FTs: two-man light tanks much smaller, though not much faster, than the British machines.[14] Soon, however, all the Allies were in a roughly equal condition.

The inevitable peacetime retrenchment took two forms. One was economic; units were disbanded, manpower levels and budgets were reduced as far as practicable, and the remaining forces were expected to use up inventories acquired on wartime scales at sharply limited peacetime rates of expenditure. This was to be expected, but it worked a hardship on the capital- and innovation-intensive tank forces. Consider, for instance, that the British got four generations of heavy tanks to the front in the twenty-seven months separating Bapaume in September 1916 and the Armistice, and none for eighteen years thereafter.[15]

Perhaps a greater hardship was worked by institutional, rather than fiscal or material, retrenchment. The tank organizations of the Great War were comparatively ad hoc wartime entities, and their leaders mostly holders of wartime ranks. Reverting to peacetime establishments and affiliations, they were in no political position to make their case with the necessary vigor as their concentrated expertise was demobilized and dispersed. Nor were their seniors in much of a position to hear it; they faced a return to prewar commitments and to a certain extent prewar force structures, all on near-prewar budgets. Combined with the prewar attitudes of many senior officers, this left little room for the pursuit of innovative technologies. This atmosphere of reaction—the proverbial post-Armistice desire to return to "real soldiering" described (if in exaggerated form) in memoirs and articles by such tank partisans as Fuller and B. H. Liddell Hart[16]—was accompanied by one of rationalization. In drawing lessons learned from the war, most armies concluded that trench warfare was a fluke, and that specialized equipment for its prosecution could be deemphasized and savings realized. Tanks, strongly associated with neither the cavalry, infantry, nor artillery, were generally considered to fall into this category, and thus were often considered largely dispensable.

The U.S. experience was like that of the other Allies. The Armistice found the Army with an inventory of 1,000 license-built Renault FTs and 100 British-derived Mark VIII tanks. These machines formed essentially the entire armored vehicle fleet until 1930.[17] Under the terms of the National Defense Act of 1920, these machines, regarded as infantry support weapons, belonged to the Infantry branch, which held a legal monopoly on tanks in the Army. Equitably enough, the Cavalry held a parallel monopoly on the wheeled armored cars associated with reconnaissance duties; sporadically, they operated one company's worth.

These machines soldiered on in auxiliary roles consonant with their age and dignity until the summer of 1927, when the status quo was shaken by events abroad. The year before, after prolonged lobbying led from within the Army by Fuller and from without by Liddell Hart, and after heated intramural arguments culminating in Fuller's resignation—later withdrawn—the British Army had fielded an experimental mechanized force of roughly brigade size. Its maneuvers, held on Salisbury Plain, drew wide attention. One observer was U.S. Secretary of War Dwight Davis, who directed the Chief of Staff, General Charles P. Summerall, to emulate the British effort. Summerall in turn sent his G-3 (Operations and Training) officer the simple order: "Organize a mechanized force." The resulting formation based on single battalions of light tanks, heavy tanks, truck-mounted infantry, and artillery, all equipment of wartime vintage, took the field at Fort Meade in July 1928. Breakdowns and shortages of the force's aged equipment led to termination of the initial experiment by September. The idea of mechanization gained new support and publicity, but little was learned from from the exercise.[18] Within two years, however, a more permanent force with newer vehicles was raised at Fort Eustis, Virginia. The new formation again faltered for lack of financial support; nonetheless, it laid the foundation for further sporadic work through the 1930s.

These early mechanized forces, as well as those of France, the Soviet Union, and one being developed in secret by Germany, have all been thoroughly studied.[19] For the purpose of this work, there is one important thing to consider: in each case, the forces' infantry components were standard rifle and machine gun units conveyed in standard, unarmored trucks: "motorized" rather than "mechanized," in modern terms. This gave the infantry comparable, if not superior, operational mobility to the tracked components; still, their tactical mobility—the ability to move with the tanks in the face of fire and adverse terrain—was no greater than it had been in 1917, an obvious shortcoming quickly confirmed by maneuvers and war games.

A variety of technological advances made this state of affairs more serious than it had been previously. The new generation of tanks could now outpace accompanying infantry: even machines in development at the end of the war had cross-country speeds over 10mph; they could thus leave advancing infantry behind even in the absence of enemy fire. At the same time, the prospect of the proliferation of antitank weapons made it compelling to move at such speeds to reduce vulnerability. The lee of a tank was less useful as shelter as well. More infantry mortars were being fielded and better artillery fire control grew in prospect, so that high explosive rounds might assail advancing infantry from almost any direction by means of indirect fire. In any case, newer tanks were smaller than the trench-spanning behemoths of World War I.

The lack of vehicles appropriate to the infantry components of mechanized forces grew less from any institutional obtuseness than from the economics of the early mechanized forces. Much as navies in the age of sail built ships of the line in their shipyards in preference to smaller vessels, which could be more quickly and more widely constructed, the embryonic mechanized forces of the major powers concentrated on the development and procurement of the tank. Like the ship of the line, the tank was the most powerful element and the one requiring the most developed industrial base and pool of design expertise.

That is not to dismiss institutional obtuseness as a factor; there was a good deal of it, much of it, oddly enough, deriving from the visionary ideas of such theorists as J. F. C. Fuller and B. H. Liddell Hart. The former, in particular, saw little use for infantry on the modern battlefield. In his words of 1932: "To combine tanks and infantry is tantamount to yoking a tractor to a draught horse. To ask them to operate together under fire is absurd."[20] In Fuller's dispensation, infantry's role was almost peripheral to the land battle, employed as "field police" for mop-up and occupation duties, and as light infantry for operations in terrain unsuitable for tanks.[21] The infighting and assault roles were to be taken up by the tanks themselves. Liddell Hart, at least, foresaw a role for "tank marines," themselves mechanized, clearing obstacles and holding ground.

Different armies came to grips with the problem in different ways. All the powers attempted to multiply the presence of armor by emphasizing the machine gun-armed light tank as an infantry surrogate, skirmishing before the main force: a "light infantry tank," in effect. The Soviets, with access to much manpower and little materiel, largely ignored the problem otherwise, concentrating on procuring tanks to the exclusion of other types of armored vehicles. As for the Germans and Americans, they simply noted the existence of the technological problems of providing close support to their tank forces and trusted their advanced automotive industries to provide when better or needier days came. The British, on the other hand, attempted to design their armored vehicles to do without close infantry support through the use of tanks with multiple machine gun turrets, such as the abortive Vickers "Independent," the A.6 "Sixteen-tonner," and the somewhat similar A.9, and through the use of subsidiary, lightly armored "tankettes." They were driven in this course less by the vision of the likes of Fuller than by the institutional memories of their armored vehicle operators. The cavalry regiments had their own independent ethos, while the rump of the Tank Corps had the memory of machines such as the Mark V, capable of engaging up to six different targets at once, disposed around the clock.

Unfortunately, transferring this idea from the sponsoned layout of British tanks of Great War vintage, characterized by hull-mounted weapons of limited traverse, to that of more modern machines carrying their armament in turrets was problematic. Providing multiple turrets for comprehensive coverage made for complex tanks that were bulky, cramped, or even both at once. At the same time, such an approach led to increased vulnerability to projectiles: the required openings for the additional turret rings sacrificed structural integrity in the hull; meanwhile the mechanical needs of the turrets themselves—their sighting, access, and ventilation arrangements—added additional embrasures and shot traps for incoming projectiles. With their complexity, these machines tended toward mechanical unreliability, and (much worse for procurement authorities) greater expense. The three-turreted A.6 of 1928, for instance, was turned down almost entirely due to its projected price tag of £16,000.[22]

In short, the tank was developed in World War I as an infantry support weapon. Even in that limited role, however, it was apparent that infantry trained and equipped to cooperate with the new machines was necessary; in particular, that it was useful for such troops to have specialized transport. The straitened financial circumstances of the interwar period, however, combined with the ideological conflict over the nature of the first mechanized formations, to stultify this line of development.

NOTES

1. A. J. Smithers, *Cambrai: The First Great Tank Battle, 1917* (London: Leo Cooper, 1992), 51.

2. Richard Simpkin, *Mechanized Infantry* (Oxford: Brassey's, 1980), 9–17. Simpkin recognizes the legacy of mounted infantry, dragoons, and the like, but adds that mechanized infantry owes more to the attitudes and traditions associated with light infantry units, although this can be said of almost all infantry since the adoption of the rifle as the standard individual weapon. Daniel J. Kaufman, "Organizations, Technology, and Weapons Acquisition: The Development of the Infantry Fighting Vehicle" (unpublished diss., Massachusetts Institute of Technology, 1983) follows Brigadier Simpkin's model. See the essay heading the bibliography for more on Simpkin, whose works provide much of the theoretical underpinning of this book.

3. A. J. Smithers, *Cambrai*, 49f.

4. R. Ernest Dupuy and Trevor N. Dupuy, *The Encyclopedia of Military History from 3500 B.C. to the Present* (New York: Harper and Row, 1986), 961.

5. Smithers, *Cambrai*, 97.

6. J. P. Harris, *Men, Ideas, and Tanks: British Military Thought and Armored Forces, 1903–1939* (Manchester: Manchester University Press, 1995), 108. Harris contends that Cambrai is of more importance for the British use of these techniques than of armor.

7. Dupuy and Dupuy, *Encyclopedia of Military History*, 970.

8. D. S. Graham, "The Ascendancy of Firepower," in J. A. English, J. Addicott, and P. J. Kramers, eds., *The Mechanized Battlefield* (Washington, DC: Pergamon-Brassey's, 1985), 70 n. 3.

9. Smithers, *Cambrai*, 51.

10. B. H. Liddell Hart, *The Tanks* (London: Cassell, 1959), I: 156.

11 Smithers, *Cambrai*, 50.

12. Liddell Hart, *The Tanks*, I: 179f.

13. *Ibid.*, I: 124.

14. Cf. I. B. Holley, Jr. *Ideas and Weapons: Exploitation of the Aerial Weapon by the United States during World War I* (New Haven CT: Yale University Press, 1953), for a similar case with aviation, the more so since prodigal application of the Liberty engine, in this case to the Mark VIII tank, was also a stumbling block.

15. Liddell Hart, *The Tanks*, I: 368.

16. A brief summary of this retrenchment in the British army can be found in Liddell Hart, *The Tanks*, Volume I, Part II, chapter I; an overview of Liddell Hart and Fuller's publications on the matter can be found in chapters 4 and 5 of A. J. Trythall, *"Boney" Fuller: Soldier, Strategist, and Writer, 1878–1966* (New Brunswick NJ, Rutgers University Press, 1977).

17. Mildred Hanson Gillie, *Forging the Thunderbolt: A History of the Development of the Armored Force* (Harrisburg, PA: Military History Publishing Company, 1947), 17.

18. *Ibid.*, 20f.

19. A recent case-by-case overview may be found in the early chapters of J. P. Harris and F. P. Toase, eds., *Armoured Warfare* (New York: St. Martin's, 1990).

20. J. F. C. Fuller, *Armored Warfare* (Harrisburg PA: Military Service Publishing Company, 1943), 16. In this revision of earlier work, Fuller qualified this statement to the effect that: "When I wrote this I had in mind infantry as commonly understood—namely, as footsoldiers whose raison d'etre is to fight their like. Motorized infantry equipped with anti-tank weapons are not infantry, they are anti-tank foot" (*ibid.*).

21. *Ibid.*, 18.

22. Liddell Hart, *The Tanks*, I: 368.

Chapter 2

Armored Infantry in World War II

> My first moment of action was when I was marker tank in the Tobruk breakout and a very brave German jumped on my back flaps armed with a Molotov cocktail and a crowbar. We had a doubly lucky escape. My troop corporal was covering but his gunner pressed the wrong button and a two-pounder shot went straight through the German and nicked the armor by the small of my back. . . . the tank being a diesel one, we managed to get the engine fire out. I must admit that ever since then I've suffered a certain amount of "infantry terror."
>
> —Richard Simpkin[1]

For the U.S. Army, two decades of restricted means ended with the 1930s. U.S. rearmament came with relative abruptness as world tensions rose with the rearmament of Nazi Germany, New Deal works programs subsidized new equipment, and Great War stockpiles finally began to dwindle. As part of the general reequipment of the armed forces, a new, permanent Armored Force was established, including a specialized infantry component. As discussed above, these troops were originally regular infantry mounted in trucks, but this arrangement lacked the armor protection and cross-country mobility possessed by the tank-based components they were to work with. An early proposal was to mount riflemen in lightly armored scout cars of the type already used by mechanized cavalry units, but by 1941, a better solution was at hand in the form of the M2 half-track and its lengthened derivative, the M3 half-track. These machines were, in essence, trucks with tracked running gear replacing the rear wheels, a French development the Army had experimented with since the 1920s. The M3 could accommodate an infantry rifle squad of eleven men and their equipment in its rear compartment, in addition to a crew of two forward, although usually only a driver was assigned. Although the new vehicle provided less armor and more limited cross-country mobility than the tank, what it provided was a considerable step forward.

Once doctrine was evolved to accommodate its characteristics, the half-track was the ingredient that turned motorized infantry into armored infantry, capable of much closer cooperation with tanks. The U.S. Army, in fact, was unique in its ability to so equip all the infantry components of its armored forces. While thinly armored compared to contemporary tanks, it was nonetheless resistant to most small arms of the time, although it lacked overhead cover. It was, in short, typical of the U.S. armamentarium on the eve of World War II: arguably inferior in mobility and protection to some foreign equipment, such as the German Sd.Kfz. 251[2] three-quarter-tracked carrier, but still nonetheless serviceable, and available in tens of thousands—great enough numbers to fully equip the entire armored infantry force. Its enemy counterpart equipped only 26 out of 226 *Panzergrenadier* battalions at the peak of its availability.[3]

As it turned out, the easy availability of the half-track as an infantry vehicle was most fortunate. In common with the armored forces of the other belligerents, U.S. armored divisions had too little organic infantry for most of their missions, only a regiment of two armored infantry battalions available organically to support eight tank battalions. As a result, relatively minor obstructions—field fortifications, close terrain such as forests, or urban areas—could slow the division's advance by allowing enemy infantry to close with tank forces in relative safety. In the absence of sufficient armored infantry to clear such obstacles, it was necessary to bring up additional line infantry. As such troops lacked the transport to keep up with armored forces, rates of advance were slowed in any case. They could be stopped completely if enemy fire forced the accompanying infantry to take cover. The only alternative, that of doing without infantry support, was unacceptable. The British anecdote at the head of this chapter is typical of the consequences of operating tanks unaccompanied in the wrong conditions, aside perhaps from the outcome.

For all the armored divisions of the combatants, then, the key to success was the proper balance of tanks and infantry. As with other formations based on interwar ideas, the first U.S. armored divisional organization, derived from that of the 1940 German *Panzer* division,[4] was intended for pursuit and exploitation, rather than the decisive engagement of opposing armored forces. The latter were to be left to specialized tank destroyers, lightly armored but highly mobile antitank gun carriers. The armored division was thus intended to face a disorganized opponent, maneuvering and reconnoitering its way around resistance. To this end, light tanks were a major component, in fact preponderant in the original establishment.[5] By March 1942, however, prior to the armored division's North African combat debut, the organization had been revised in the light of foreign wartime experience, which had shown the weaknesses of light tanks as the mainstay of armored forces. The two armored regiments in the division were now reconstituted with two medium and one light tank battalions for a total of 232 medium and 158 light tanks The armored infantry regiment meanwhile gained a third battalion.[6] All these regimental structures were essentially administrative. Tactical control was vested in two "combat commands" to which units were attached as dictated by conditions.

This "heavy" armored divisional structure, however, was not definitive; early combat experience combined with logistical considerations to produce a radical reorganization. Medium tanks had shown themselves to be even more useful relative to light ones than had been thought by prewar force planners. The superimposition of the traditional regimental organization onto the combat command system had shown itself to be unwieldy, increasing the number of headquarters in the division while leaving some idle. Finally, even more armored infantry was seen to be necessary for operations. Simply attaching line infantry as needed, in line with the "pooling" philosophy that drove Army force design, was problematic. Such units were not equipped or trained to work efficiently with armor. In particular, the standard U.S. "triangular" infantry divisions, organized with economy of manpower and strategic mobility in mind, lacked the organic transport to move its maneuver elements save by disrupting its artillery and logistic operations. Plans to raise "armorized" infantry divisions, their line units mounted in half-tracks, had fallen by the wayside due to the logistical penalties involved.[7]

These same two factors, strategic mobility—especially trans-Atlantic shipping space—and manpower, shaped a radical revision of the armored division in September 1943. The chief of Army Ground Forces, Lieutenant General Lesley J. McNair, with these two considerations always foremost in his mind when approving force designs, had never been happy with the armored division's organization. Protesting its 14,620 men and thousands of tons of vehicles, he called it "an amazing picture of unjustified extravagance" in the face of the problems of manning and supplying the infantry divisions needed for projected operations in Europe.[8] As a result, the need for parity of armored infantry with tanks in the armored division was met not by adding infantry, but by paring tanks from the formation. One of the four medium tank battalions was deleted; light tank units were eliminated entirely, a company of light tanks being added to each of the three remaining medium battalions for close reconnaissance. The resulting division was thus built around three tank, three armored infantry, and three armored artillery battalions. This complement was small enough to be controlled tactically by three combat commands—a third, smaller command, intended to control reserves, having been added—with no other headquarters between division and battalion. The elimination of excess headquarters made for striking economies: the new "light" armored division of 10,937 men with 186 medium and 77 light tanks, while keeping the same armored infantry and artillery strength, fielded about the same fraction of tanks per capita as the older formation. In addition, a greater proportion of these were medium tanks, with concomitantly greater combat power.[9]

While the armored division and the tank battalion went through radical organizational shifts, the armored infantry battalion's establishment stayed relatively static. The 1944 organization is representative: a headquarters company, including mortar, reconnaissance, and assault gun platoons; and three armored infantry companies, each of three rifle platoons and an antitank platoon. Each rifle platoon, in turn, was built around three rifle squads and mounted in five half-tracks. Each squad had its own vehicle, one of them sharing it with the platoon headquarters; the additional two vehicles carried a light machine gun squad and a 60mm mortar squad.[10] The mortar was attached at platoon level, rather than being part of a company mortar section, as in regular infantry units, in view of the more dispersed nature of armored infantry operations and the quicker response necessary. The armored infantry squads themselves likewise were derived from conventional rifle squads. They were, however, modified to fit their mechanized environment. An additional man was added to drive the half-track, bringing the total strength to twelve men. Also, the squad's weapons were somewhat different. Until 1945, all squad members save the driver were armed with the M1 semiautomatic rifle; the Browning Automatic Rifle (BAR) present in each regular infantry squad was replaced as the base of fire by the heavier .30 caliber machine gun normally mounted on the vehicle, which could be removed and mounted on a tripod for dismounted operations. In addition, each squad was issued a bazooka when that weapon became available, thus giving it an independent, if limited, antitank capability.[11] Armored infantry was generally expected to engage in combat dismounted. At the same time, doctrine urged that armored infantry move "forward in vehicles until forced by enemy fire, or unfavorable terrain, to dismount."[12] Vehicle-mounted weapons were used for supporting fires, especially since armor infantry lacked the BAR as a squad weapon, and in emergencies, the half-tracks themselves could maneuver in support of the dismounted infantry, as at Medjez el Bab, Tunisia, on 5 December 1942, when Company A, 6th Armored Infantry Regiment used its half-tracks in a series of mounted counterattacks against German infantry.[13]

The armored infantry battalion thus had an impressive degree of firepower and mobility. As one contemporary observer at the Army Command and General Staff School noted, the half-track-equipped infantry squad had greater capabilities than many World War I tanks.[14] The armored infantry battalion was therefore considered capable of a wide variety of missions, most though not all connected with the support of tank forces. By 1944, under official doctrine, these included responsibilities to:

a. Follow a tank attack to wipe out remaining enemy resistance.
b. Seize and hold terrain gained by tanks.
c. Attack to seize terrain favorable for a tank attack.
d. Form, in conjunction with artillery and tank destroyers, a base of fire for a tank attack.
e. Attack in conjuction with tanks.
f. Clear lanes through minefields alone or in conjunction with engineers.
g. Protect tank units in bivouac, on the march, in assembly area, and at rallying

points.
h. Force a river crossing.
i. Seize a bridgehead.
j. Establish and reduce obstacles.
k. Attack or defend towns.
l. Organize and defend a position.
m. Perform reconnaissance and counterreconnaissance.[15]

The battalion seldom if ever operated in its pure form, though. In combat, the armored infantry battalion headquarters company, like those of divisional tank battalions, normally acted as the nucleus of a task force formed by cross-attachment: the combat command would trade companies among its assigned tank and armored infantry battalions as its mission dictated, supplementing them with other forces under their organization such as attached tank destroyers or combat engineers. This process continued down the tables of organization, sometimes to even the lowest levels, as seen in this example from the Fifth Armored Division:

> One tank battalion and one infantry battalion were assigned to each combat command, but the cooperation between tanks and infantry go farther than that. The "A" tank company and an infantry company were paired off. Likewise, "B" and "C" tank and infantry companies. Within the companies, each tank crew of 5 men were paired off with its own infantry squad of 12. The final result of the marriage was a Sherman tank, a halftrack, and 17 men who were to eat, sleep, and fight together.[16]

The half-track was fundamental to all the foregoing. Available in considerable numbers due to mass production, it was ubiquitous within the armored division, and was adaptable to numerous roles beyond the basic one of armored squad carrier. It served variously as a tank destroyer, mortar carrier, anti-aircraft vehicle, armored command post, gun tractor, and maintenance vehicle, among other applications. Combat experience in the armor-intensive North African and European Theaters of Operation, however, made some drawbacks to the vehicle evident. Although its cross-country mobility, both in terms of raw speed and the ability to negotiate obstacles, was superior to that of the truck, it was decidedly inferior to that of the tanks it worked with, placing strictures on operations as a result. Similarly, while its quarter-inch steel armor was generous for its prewar design, it no longer matched the threat at hand. Although the half-track's side protection was reckoned even in 1945 as resistant to hits from ".30-caliber ball ammunition at all except extremely close ranges; from .30-caliber A[rmor-]P[iercing] ammunition at ranges over 300 yards; and from .50-caliber ammunition at ranges over 1,000 yards,"[17] the proliferation of heavy weapons on the battlefield made even this level of protection insufficient. In addition, the vehicle, showing its scout-car antecedents, was open-topped; it was thus highly vulnerable to artillery and mortar rounds bursting overhead, when artillery was the very threat it was most necessary to protect armored infantry from if they were to cooperate closely with tanks.[18]

A number of expedients were used to cope with these drawbacks. The most obvious solution to the mobility differential was for infantry to ride on the tanks, in the fashion made famous on the Eastern Front by the Red Army. With improvised handholds and careful drill, each medium tank could carry five or six men—a half squad—behind the turret on its engine deck,[19] but this exposed the riders to grave risks in the face of any artillery threat. The tank also provided an expedient solution to the protection differential as well, especially among British Commonwealth forces. Redundant tanks with their turrrets removed called "Kangaroos," notably the otherwise unsuccessful Canadian Ram (a Sherman derivative), could accommodate a rifle squad with impressive protection from direct fire, although except in the case of a few especially-sophisticated conversions the problem of overhead protection remained. M7 self-propelled 105mm howitzer hulls, minus their ordnance, served in similar roles.[20]

More refined attempts were also made to address the problems of the half-track as well. Stateside experiments were conducted with a variety of "armored utility vehicles" derived from the M18 tank destroyer. The first of these to reach service, the M39, was originally open-topped, thus sharing the disadvantages inherent in the half-track's layout. Furthermore, it was quite cramped when used as a squad carrier. As a result, the M39 was used mainly as a tractor for antitank guns.[21] A derivative, however, the M44, a shedlike armored box on tracks capable of holding half a platoon, remedied these problems although its large, thinly-armored silhouette was a problem. As with the British Mark IX of 1918, however, the war ended before it reached service, and only a few were procured before postwar retrenchment caught up with it.

By the end of the war, then, the armored formations of the major belligerents had steadily increased the relative strength of their infantry components. These tended to evolve specialized organizations and doctrines to meet their missions. At the same time, where it was industrially feasible, it was found useful to equip such units with specialized vehicles to enhance their effectiveness. All such vehicles, however, were more or less makeshifts, and thus regarded as unsatisfactory, especially as a more intense battlefield environment evolved.

NOTES

1. "Men Against Tanks," in John A. English, J. A. Addicott, and P. J. Kramers, eds., *The Mechanized Battlefield: A Tactical Analysis* (Washington, DC: Pergamon-Brassey's, 1985), 139.

2. *Sonderdienstkraftfahrzeug*, special purpose vehicle.

3. Simpkin, *Mechanized Infantry*, 22.

4. Kent Roberts Greenfield, Robert R. Palmer, and Bell I. Wiley, *The Organization of Ground Combat Troops* (Washington, DC: USGPO, 1947), 322.

5. Gillie, *Forging the Thunderbolt*, 216.

6. Jonathan M. House, *Toward Combined Arms Warfare* (Ft. Leavenworth, KS: U.S. Army Command and General Staff College, 1984), 109.

7. Greenfield et al., *Organization of Ground Combat Troops,* 324–27.

8. Memo (S) of General McNair (dictated by General McNair, signed by General J. G. Christiansen) for DCoSUSA, 20 Jul 43, sub: Comments on Major General Harmon's Rpt. 319.1/13 (NATO) (S), in Greenfield et al., *Organization of Ground Combat Troops,* 322.

9. House, *Toward Combined Arms Warfare*, 109.

10. *Armored Infantry Company, War Department Field Manual FM 17-40* (Washington, DC: USGPO, 1944), 2. Hereafter "*FM 17-40.*"

11. Virgil Ney, *The Evolution of the Armored Infantry Rifle Squad* (Ft. Monroe, VA: U.S. Army Combat Developments Command, 1965), 79–82.

12. *FM 17-40*, 3.

13. Paul McDonald Robinett, *Armor Command* (Washington, DC: McGregor and Werner, 1958), 89.

14. William L. Roberts, "Employment of Armored Infantry of an Armored Division," *Military Review* 23 (May 1943): 33.

15. *Armored Infantry Battalion: War Department Field Manual FM 17-42* (Washington, DC: USGPO, 1944), 2f.

16. *The Victory Division in Europe, A Unit History of the 5th Armored Division* (Goth [*sic*], Germany: Englehurd and Associates, 1945), 4; quoted in Dale K. Brudvig, "A Proposed Divisional Organizational Concept," in *Selected Readings in the Development of Combat Divisions* (Ft. Leavenworth, KS: U.S. Army Command and General Staff College, 1973), 2-104.

17. J. B. Evans, "Armored Infantry," *Military Review* 24 (March 1945): 40.

18. Daniel J. Kaufman, "Organizations, Technology, and Weapons Acquisition: The Development of the Infantry Fighting Vehicle" (unpublished diss., MIT, 1983) also points out that the open top made the half-track vulnerable to grenades and other explosive attacks when moving through towns (98), although it should be noted that this is just when armored infantrymen would be expected to be dismounted.

19. *FM 17-40*, 164–76.

20. R. P. Hunnicutt, *Sherman* (Belmont, CA: Taurus, 1978), 486f.

21. R. P. Hunnicutt, *Stuart: A History of the American Light Tank, Volume I* (Novato, CA: Presidio, 1992), 285–88. An armored top was devised for the M39, but it made entrance and exit from the vehicle difficult, and further aggravated the space problem.

Chapter 3

Infantry with Tanks: From Armored Infantry to Mechanized Infantry, 1945–1965

> A tanker halts between moves; an infantryman moves between locations. This simple statement depicts two ways of life and thought almost as different as those of a sailor and a landsman. It also explains why most tank-infantry cooperation breaks down before it has started.
>
> —Richard Simpkin[1]

Although the point can easily be carried too far, the experience of land warfare in the African and European theaters of World War II may be broadly described as a conceptual shift from massed infantry supported by tanks to a roughly equal partnership of combined arms, as opposed to the war of massed tanks supported by infantry envisioned by the interwar theorists of armored warfare, a situation epitomized by the drift toward balance in the proportions of infantry to armor in the British, German, and U.S. armored divisions over the course of the war.

No sooner had this shift been made, however, than the situation changed again. Land warfare as waged in the Western tradition was overshadowed by the advent of nuclear weapons and their intellectual and material demands. At the same time, its conventions were challenged by the outbreak of warfare outside the developed world.

These twin upheavals had profound effects on the postwar U.S. Army. The newly spawned Air Force claimed the lion's share of defense resources under the nuclear dispensation, with the Navy fairly close behind; the Army was firmly in third place. The Army therefore had less buying power while it had to exercise what it did have in an environment heavily influenced by the technology-intensive values of the other services. This predicament tended to impose extra complications on its procurement decisions, decisions already rendered complex by the existence of simultaneous demands to fight the Soviet Union on a potentially nuclear battlefield in Europe, while being ready to intervene anywhere else in the world at the same time.

Demobilization hit the Army hard. On 31 May 1945 the service's strength was 8,291,336; twenty-five months later, it was 989,664. The strength of the Infantry branch alone went from 1,782,832 to 126,121. As a result of the nature of the postwar force structure, geared as it was to occupation duties, armored infantry was proportionally an even smaller element. Given that only one of the twelve divisions remaining active after demobilization was armored, the subset of armored infantry, tied as it was to the armored division, became an even smaller fraction of the infantry as a whole.[2] Hiving off the former Army Air Forces as an independent service had a further impact, the more so since the Air Force, with its technological focus and monopoly on the nuclear deterrent, gained priority in the allocation of resources.

Although it faced theoretical worldwide employment, the main commitment of the postwar Army was in Europe as the keystone of the newly formed North Atlantic Treaty Organization (NATO), facing much larger Soviet ground forces under the nuclear "umbrella" provided by the now-separate Air Force: well under, given the U.S. nuclear monopoly in the immediate postwar era; under such conditions, the Army's combat role often seemed almost an afterthought. In the face of these conditions—out of the limelight, constrained in resources, and planning for campaigns in a theater hard by its recent victorious battlefields—the Army's doctrine, organization, and equipment tended to be evolutionary developments from those of wartime.

This was particularly true for the elements of the erstwhile Armored Force. Wartime experience had forced a reevaluation of many key features of doctrine, especially as regarded McNair's ideas on the relative value of economy versus capability on the modern battlefield. His notion that relatively lightly equipped and homogeneous infantry divisions could be augmented by heavy or specialized equipment pooled under the control of higher formations, in particular, was deemphasized, if not abandoned. Because this was the underlying philosophy behind the organization and equipment of the infantry division, these formations changed fairly significantly in character. Tank units, among others, became organic to the standard infantry division for the first time, as the division departed from its former ideal of lightness for one of durability in the face of sophisticated threats.

The armored division also grew heavier. However, it changed less in concept, having gone through so many organizational changes during wartime. The principle of balanced proportions of tank and infantry remained; so did that of tactical control by combat command headquarters rather than regiments or brigades. On the other hand, various units habitually attached to the armored division were now made organic to it, such as battalions of 155mm self-propelled howitzer and light antiaircraft artillery. Further, in view of the wartime failure of the tank destroyer doctrine, which envisioned specialized mobile antitank assets rather than tanks of the armored division themselves engaging enemy armored formations, a fourth tank battalion, equipped with heavy tanks, took over the antitank defense of the formation. To balance it, a fourth armored infantry battalion was added to the tables of organization and equipment.[3]

In essence, the armored infantry battalions were similar to those of wartime, as were the individual and crew-served weapons. The major equipment of the armored infantry batallions themselves, however, was subject to change. Wartime experience had shown that though the half-track was a vast improvement over trucks and other wheeled, unarmored vehicles as transport for infantry supporting tanks, a full-tracked vehicle was needed to keep up in cross-country movement. Furthermore, given the official mission of the armored infantry carrier vehicle—carriage of troops up to, but not into, battle—the protective arrangements of the half-track were inefficient. An expedient development, the half-track had been derived from a prewar cavalry scout car, a reconnaissance vehicle that had originally sported an open top for better visibility. The open top provided a quick means of egress and allowed squad weapons to be brought to bear when mounted combat developed in spite of doctrinal prescription, but it left the squad dangerously bereft of overhead cover in a battlefield environment where air-bursting shells had been shown to produce a large number of casualties. This threat became much more serious as VT ("variable-time" or proximity)-fuzed rounds became widespread instead of the closely held U.S. secret they had been in wartime. These radar-fuzed shells could be fired more quickly and efficiently than previous munitions, not being reliant on time-fuzing calculations; they nonetheless posessed the greater lethal radius of air-bursting munitions against unarmored targets. Given that armored infantry vehicles were intended for use behind the line of contact where artillery was the great threat, this situation was unacceptable. Similarly, the possibility of operations on a nuclear battlefield, with its accompanying blast, heat, and radiation threats posed problems for the conduct of the sort of combined-arms operations the Army had come to embrace.

The first design attempting to address these deficiencies was in hand, though not in service, at war's end (see chapter 2). The M44 Armored Utility Vehicle was, in essence, a thinly armored box on tracks, sharing some mechanical components with the new M41 light tank. With a capacity of twenty-seven men and a profile often described as "shedlike," it quickly proved unsuitable as a tactical vehicle, presenting too large a target and endangering half a platoon while doing so.

In recognition of these drawbacks, efforts got underway immediately after the war to design and procure a successor to the M44. In September 1945, Army Ground Forces established a requirement for the new vehicle, which was then developed by the Ordnance Department as the T18 armored utility vehicle. Originally conceived as an elaborate vehicle equipped with a pair of remote-controlled machine gun turrets, the T18 underwent a variety of simplifications over the course of its development, emerging from the process with the new designation "Vehicle, Armored Infantry, Full-Track, M75" in December 1952.[4]

The new vehicle kept a number of the features of the M44, notably overhead protection and rear exit for the troops carried. It was, however, considerably smaller. Sized to carry a twelve-man armored infantry squad, the M75 represented a significant gain over the half-track and its derivatives in mobility, and its overhead cover gave increased protection from anticipated threats.

The M75 was not without drawbacks, however. It was lighter and smaller than the M44, yet it was still a bulky vehicle, and its 21 tons represented a greater weight than infantry units were used to accommodating in their movement and loading plans. Its mobility was decidedly superior to that of the half-track, yet it was still notably inferior to that of the M47 tanks it was supposed to work with, due to its lower power-to-weight ratio and shorter length,[5] which made it more sensitive to horizontal obstacles, such as ditches. Furthermore, overhead protection had been obtained at a price in habitability; the passenger compartment was notably ill-ventilated and dark, and the only weapon that could be brought to bear while the squad was mounted was the .50 caliber machine gun mounted at the commander's station as an antiaircraft weapon. Egress was limited to two small doors at the rear, through which ten men were expected to disembark in twelve seconds.[6]

In 1952, the M75 was nearly withdrawn from service.[7] Most of its drawbacks could be defended as trade-offs made in conformity with doctrine, but the machine's greatest flaw was its cost. Priced as high as $132,000[8] largely due to its employment of tank components and structural design practice, the M75 was regarded as a strain on the finances of the Army. The advent of the Korean War might have boosted the vehicle's popularity, but due to the geography of the theater, no armored infantry units fought there.[9] As a result, regular infantry forces were raised in greater strength than armored ones. A number of M75s, however, were used as armored supply vehicles and ambulances.[10]

To address the problems of the M75, especially that of cost, a new vehicle was called for in 1950 and fielded by 1954, supplanting the M75 in production.[11] The great advantage of the M59 APC was that it was much cheaper, at $28,000.[12] This savings arose mainly from the use of two commercial General Motors gasoline engines, as found in the standard 2½-ton truck, in place of the more specialized single power plant of the M75.

This economy came at a price, however. Aside from the obvious mechanical complication of twin engines, the new power plant gave about five-sixths of the horsepower of the M75's Continental gasoline engine. Given that the M59, while somewhat smaller than the M75, was slightly heavier, the result was a slower and less agile vehicle, with a top speed of 32mph versus almost 45mph for the M75. This shortfall exacerbated the mobility differential between cooperating infantry and tank units.

The M59 was not simply an exercise in value engineering, however. It addressed a number of the M75's perceived shortcomings. The vehicle was somewhat lower, thus presenting a smaller target. Squad egress was improved by the adoption of a hydraulically operated rear ramp in place of the M75's two small rear doors. The commander's exposed, ring-mounted .50 caliber machine gun was replaced early in FMC's 4,700-vehicle production run by a turreted weapon that could be fired from under armor. The new, heavier mounting was much more effective for ground combat due to its greater protection and steadiness. However, it was unpowered, and therefore limited in its speed of traverse and elevation, a tacit admission that the original antiaircraft role of the weapon was dated in the jet age. Despite this improved provision for combat, however, the infantry squad carried in the vehicle still lacked any facilities, such as firing ports, for mounted combat; it was necessary for them to dismount to employ any weapons.

In addition to its other useful characteristics, the M59 gained a limited amphibious capability. It thus actually added to the ability of infantry units to cross water obstacles, rather than throwing a strain on bridge and ferry capacity as did the M75. The M59 gained buoyancy from its boxy shape, and was propelled in the water by its tracks at slightly over 4mph. Additional longitudinal stability was provided by a plywood trim vane that unfolded from the bow of the vehicle; this became a standard feature on U.S. amphibious armored vehicles.

The M75 and M59 were relatively similar, so the doctrine for their employment was similar as well. It represented a codification of World War II practice in the European Theater of Operations: armored infantry units were expected to operate in close association with tank units. Neither type of battalion was generally expected to operate in pure form, solely on its own tables of organization and equipment, but rather as "task forces" created by armored division combat commanders trading, or "cross-attaching" companies from the units allotted them by the divisional commanders. In turn, task force companies swapped platoons to form "teams," company-sized combined-arms forces varying in composition according to mission and resources. This cross-attachment process stopped at company headquarters. In the words of one armored infantry batallion commander, Lieutenant Colonel Bergen Hovell, the limits were: "no attachments or detachments below platoon (the smallest tactical unit under an officer). No team (company) headquarters is wasted by leaving it without platoons. No platoon is attached to another except in most unusual circumstances."[13]

This intimate association threw the differential mobility, protection, and firepower of tank and armored personnel carrier into sharp contrast. The layout of the M75 and M59 made mounted combat almost impossible, and doctrine discouraged it. In Hovell's words:

> Another fallacy is that armored infantrymen somehow fight mounted. They may look impressive in a mounted charge and they may be closing in for the kill, but they are not ready to fight. Unlike tankers, armored doughboys must dismount to fight.
>
> Once the infantry dismounts, the tank's pace must match that of the foot soldier if the team is to be effective. This does not mean that the tanks constantly must creep at a

slower pace, for to do this would fail to exploit their capabilities and needlessly expose them. Instead, they must move quickly, by leaps and bounds, adjusting their over-all rate to the foot soldier's progress. They may initially support the infantry by firing from defilade, then charging through or around the infantry to seize the objective. Here the significant fact is that the tanks must remain within supporting distance of the infantry. Even when the dismounted operation is completed, the tanks must remain while the armored infantry reassembles and remounts. Again, the dismounted doughboy sets the pace for the armor-infantry team. [14]

This inability to fight mounted, whatever virtue was made of the necessity, was a particular problem: it meant slowing the rate of advance of armored units as well as limiting the benefit armored infantry could derive from their carriers. Until dismounting, as one observer noted, "The driver and the squad leader are the only two persons who can see where they are going in a mounted attack. The remainder of the squad sit and wait."[15] Some idea of the tensions involved may be seen by contrasting the above passage with this excerpt from a contemporary field manual:

In the attack, the speed, mobility, and armor protection of the armored personnel carrier must be utilized to the utmost. This is best accomplished when tanks and armored infantry are employed together, each supporting the actions of the other. In order to make use of the common combat characteristics of tanks and armored infantry, armored infantry should remain in their carriers as long as possible so that —

(1) Elements of the attacking force of tanks and armored infantry can move forward at about the same speed.
(2) The battlefield mobility of both elements of the attacking tank-infantry force is retained.
(3) Casualties in areas swept by small-arms, mortar, and artillery fire are reduced.
(4) Both tanks and armored infantry can move forward supported by artillery air bursts.
(5) A degree of protection is afforded against the effects of atomic weapons.
(6) The energy of the armored infantry is conserved so that they are able to fight effectively when needed.[16]

As a result, despite differentials in mobility and protection, armored personnel carriers tended to be used in close conjunction in tank-infantry teams, the infantry vehicles following as close as five yards behind tanks in training for mounted assaults. This close association placed the carried squads at considerable risk even as their vehicles' lack of facilities for mounted combat kept them from bringing any combat power of their own to bear until they had disembarked from their carriers.[17]

With the advent of the M59, the Army had procured an affordable and useful armored personnel carrier, yet it still had drawbacks. Aside from the commander and driver, the squad had no means of observation, and thus required valuable time to orient themselves once they dismounted for combat. Size and weight were still issues, as was mobility, especially relative to that of the tank. As one user—a retired senior NCO responding to a retrospective 1981 article in *Infantry* by a senior engineer of the Army Tank-Automotive Command (TACOM)[18]—commented:

> The author states that the M75 had a top speed of 43mph, but it was not uncommon in Germany for the [M]75 to hit 60mph or more. . . . There was no way the M59 could keep up with it. The M75 could keep up with the M48 easily . . . . When the snow flew, off came the track pads, and with grouzers [*sic*] only, even the M48s couldn't stay with us.
>
> The main problem with the M75 was that it was nose-heavy. A tail-heavy tank could "float" over a small obstacle, where we would have to slow down and head into it if we did not want to stand on our nose. I still have a scar on my head from bouncing off the overhead.
>
> The article says that the M59 would top out at 32mph. It might if it was new, if its motors were perfectly tuned, and if it had a tail wind. That monster was underpowered and should have had an automotive engineer for each vehicle to maintain it.
>
> As for keeping up with tanks, forget it, especially cross country and uphill.[19]

The question of tactical mobility has been touched on, especially as regards the gap between the cross-country speed of tank and armored personnel carrier, but strategic mobility was a problem as well. The United States had lost its nuclear monopoly with the first Soviet bomb tests in 1949; subsequent developments, in any event, suggested that nuclear weapons were not as applicable to military and diplomatic problems as had been originally envisioned. As a result, the prospect of committing U.S. ground forces in times of crisis thus became much more likely. The perceived usefulness of doing so in a hurry made air transportability a desideratum; the carrying capacity and loading gauge of transport aircraft thus became as critical to armored U.S. vehicle design of the period as considerations of shipping space had to vehicle and force design in World War II.

To this end, in 1956, the Ordnance Department's Tank-Automotive Command (OTAC) established a program for a new APC. It had been studying such a machine since 1954, as the strategic-mobility problem rose in importance, envisioning a family of twenty-six modular vehicles on a single chassis.[20] The new machine was to retain the level of protection and amphibious capability of the M59, but with improved cross-country mobility and the ability to be airlifted; a weight ceiling of 16,000 pounds was therefore set. The program moved fairly quickly; a contributing factor in this was that Ordnance's vision of an extensive family of vehicles was scaled back, although multiple applications for the vehicle and its components were still envisioned. A testbed was built at OTAC, in Warren, Michigan, approximately ninety days after approval was given. Development, again by the FMC Corporation, took only forty-three months from the award of the first contract, and the company commenced production in 1960.[21]

The resulting vehicle, the M113, was superficially similar to its predecessor and stablemate, the M59. It featured a boxy armored hull riding on skirted tracks, a trim vane for swimming folded onto the sloping nose at the front, and a full-width exit ramp folded up at the rear. Like the M59, the newer vehicle was armed with a .50 caliber machine gun at the commander's station, although it reverted to an open ring mount for the weapon.

In contrast to the M59, however, the M113 was a foot lower; its internal layout was also quite different. The driver shared the front of the vehicle with a single engine, again a GM truck power plant. In place of the M59's small hatches, a large rectangular roof hatch folded back to improve lighting, ventilation, and observation for the squad under favorable conditions. Four men could stand in it, allowing them for the first time since the days of the half-track a limited capability to fight from the carrier, albeit at a considerable cost in personal vulnerability.

The most unusual feature of the M113, however, was the hull, which was made of welded aluminum armor plate, in contrast to the welded steel construction of the M59. Indeed, according to TACOM development engineer Clifford Bradley, the testbed vehicle, in fact, "was the first military vehicle in which rolled aluminum plate was used as both the structural and the ballistic material [i.e., armor]."[22]

Aluminum had a number of attractions. Perhaps the foremost one was economic: due to cancelled aircraft orders at the close of the Korean War, industrial overcapacity made the metal relatively inexpensive.[23] Cost, though, was not the sole consideration. Aluminum had several advantages as a material for armored vehicles. Pound for pound, it is more resistant to shell fragments than steel, an important factor for armored personnel carriers as they were envisioned to be used, artillery being reckoned much the greatest danger to troops in the approach march. Weight savings, though, are the foremost advantage, although the mechanism for this is not as direct as it might appear at first. Aluminum is only about a third as dense as steel, but at the same time, a given thickness is also only about a third as resistant to impact and a third as stiff structurally. Since the stiffness of a plate, though, is proportional to the cube of its thickness, an aluminum armor plate is in fact nine times as stiff as an equally resistant steel plate, allowing the omission of many of the structural stiffeners needed for a steel hull. As a result of disagreements between the Continental Army Command (CONARC, the successor to Army Ground Forces), and the Ordnance department over the merits of steel versus aluminum,[24] the M113 was tested against an otherwise identical vehicle with a steel hull, the T117. The aluminum hull showed a weight savings of 10 percent over steel. Since the remainder of the components, in particular the power train and running gear, were the same, the total weight savings came to 6 percent.[25] Given that unit cost, in turn, is sensitive to both weight and parts count, the M113 turned out even less expensive than its predecessor, the M59, at $22,000 each.[26]

These innovations in equipment were matched by organizational changes; as the first M113s left the factory, the divisions they were headed for were in the throes of reformation. Armored infantry units had operated in an essentially static divisional environment since 1947. The revolutionary, if notably unsuccessful, Pentomic reorganization of the late 1950s was never implemented for the armored divisions. When the armored personnel carrier had been introduced into the Pentomic infantry division, it had been assigned to divisional armored transport companies, rather than assigned to line units. A division's APC company was reckoned to be capable of lifting any one of the division's five battle groups, or any six of its twenty-five rifle companies, precisely analogous to a truck company or to the use of helicopters to carry air assault troops, another Army mobility initiative of the time.[27]

The ROAD (Reorganization Objective Army Divisions) initiative of the early 1960s changed this situation, however. The ROAD concept set out to standardize divisional organization as much as possible while fostering maximum flexibility in the face of evolving worldwide commitments. It established a common divisional base of services and artillery, together with three brigade headquarters analogous to the combat commands of the armored division: bare headquarters organizations with no forces permanently assigned to them. The division's maneuver battalions—between six and fifteen of them, in appropriate combinations—and other resources would then be parceled out to the brigades according to circumstances, with the brigade headquarters dictating cross-attachments into task forces and teams, again as practiced in the armored division. The ROAD developers envisioned four types of divisions using the standard base: infantry divisions, with eight infantry and two tank battalions; airborne infantry, with eight airborne infantry battalions and one of assault guns; armored, with six tank and five mechanized infantry battalions; and for the first time, mechanized infantry divisions, with seven mechanized infantry and three tank battalions.[28] In keeping with the ROAD concept, these were typical establishments rather than absolute ones, each type varying considerably in its makeup in actual service as missions and resources dictated.

The replacement of the term "armored infantry" with "mechanized infantry" should be noted. It was a side effect of a little-heralded institutional change accompanying the ROAD organizational change. Prior to ROAD, the proponency for armored infantry lay with the Armor branch; afterward, this role was taken by the Infantry.[29] This shift of proponency would eventually be problematic. It is worth remarking that the Infantry branch tended toward a monistic approach toward the various types of infantry under its aegis. Under Armor proponency, armored infantry had separate doctrinal publications in the FM 17-series,[30] as with both tanks and armored and air cavalry; articles in such publications as *Armor* carried frequent caveats to the effect that "Armored Infantry Is Different." [31] These emphasized the role of armored infantry in aiding the advance of tank-based armored formations. This focus was pronounced, and could even extend to organizational modifications, as with the Armor School's 1956 proposal to change the armored infantry platoon's light machine gun squad to a fourth rifle squad, each squad to be augmented with an additional automatic rifle.[32]

The Infantry branch, on the other hand, issued blanket doctrinal publications with catch-all titles such as that of FM 7-11: *Rifle Company, Infantry, Airborne Infantry, and Mechanized Infantry*. The thrust of this change was to give the mechanized troops less focus and orient them to the broader spectrum of infantry missions, thus rendering their training more complex and less specifically armor oriented. It also meant that the mechanized infantry units and subunits were less apt to have significant changes in organization to accomplish their missions at the cost of standardization. It was, then, with new equipment, new organization, and a new doctrinal outlook that mechanized infantry units faced their first wartime employment since 1945.

Vietnam appeared to be a most unpromising place for the employment of armored or mechanized units. An undeveloped land of forests and paddies, with the same mountains and monsoons that had hampered the use of such forces in Korea, it appeared unsuitable even in the absence of guerilla warfare conditions. This appearance, however, was somewhat misleading. An Army terrain analysis conducted in 1967, for example, determined that 61 percent of the country during the dry season and 46 percent in the wet could be traversed by tanks, whereas 65 percent was accessible to armored personnel carriers year-round; thus, though armored forces had to be used with care, and (just as in any previous conflict) as part of a larger force including other elements, especially infantry, they were by no means unemployable.[33] The Mekong Delta region—a total "no-go" area for tanks, though not in all places for armored personnel carriers, during the wet season—was the most inhospitable and the central highlands the least. By the time the study was conducted, it could be bolstered by considerable historical evidence, as well as from actual operations.

In 1962, Army conventional wisdom said armor should not be used, based on a superficial view of the environment and consideration of the opposition. Certainly there was no body of accepted Army doctrine to help in the decision. Focused on Europe as they were, Army manuals said almost nothing about jungles and little more about guerillas save in the context of rear-area security. The only real evidence was the example of the French in the previous decade, which had culminated in the destruction of *Groupement Mobile 100*, a brigade-size motorized force, by the Viet Minh in an ambush in the Central Highlands in 1954. This debacle cast serious doubts on the applicability of such forces in the country, although some pointed out that GM 100 was not properly an armored force, being essentially road-bound by the nature of its transport and with an armor element consisting of only ten light tanks.

Conventional wisdom or not, however, U.S. military assistance programs were in fact sending armored vehicles, in the form of M113 armored personnel carriers, to Vietnam. They were duly pressed into service with U.S.–advised armored cavalry regiments of the Army of the Republic of Vietnam (ARVN), where their modest weight, trafficability on available roads, and amphibious capability were expected to be useful. As early as April 1962, M113-equipped forces were organized essentially along the lines of U.S. mechanized infantry companies; their advisors expected them to fight the same way, dismounting their squads for combat. Established doctrine proved unfortunate in the context of operations in the Mekong Delta, however. In one September 1962 engagement in the Plain of Reeds, disembarking ARVN infantry quickly bogged down in marshy terrain and offered the Viet Cong (VC), who were situated in a strong defensive position, static, vulnerable targets. Mounted attacks, on the other hand, exploited the firepower of the armored personnel carriers' .50 caliber machine guns and routed the VC, who at that time lacked antitank weapons in any great quantities.[34] This situation did not last, however. The Viet Cong soon discovered the vulnerabilities of M113-mounted forces, especially the extreme vulnerability of the vehicle commander-gunners as they served their open-mounted machine guns, as well as the vehicles' susceptibility to handheld antitank weapons, and, more importantly, to mines. In much of the country, the primary threat soon evolved from gunfire to mines, to the extent that many APC-mounted troops in Vietnam, especially in the 25th Infantry Division's area, frequently rode on top of their M113s, occasionally even extending the driver's controls so that the vehicle could be driven from the roof.[35] A desire to detect ambushes through better observation and avoid some of the rigors of the tropical environment as suffered from inside a closed metal box contributed to this practice as well.[36]

Although more care was required in its use, the M113 was by no means neutralized by such threats. In both U.S. and ARVN service, it was the most common armored vehicle of any description. An essentially protean design, it sprouted numerous field modifications, many of which achieved official sanction. The M113 thus filled many roles: not just an APC, it served as a weapons carrier, command post, armored ambulance, cargo carrier, and flame-thrower vehicle, to name but a few.

The most important of these new roles, in many ways, was developed by the 11th Armored Cavalry Regiment (ACR) in 1966, building on earlier field expedients of the ARVN 2nd Armored Cavalry Regiment described above. The 11th ("Black Horse") ACR had originally relied on standard equipment for armored cavalry regiments, a mixture of tanks and M114s, which were relatively small, specialized, armored command and reconnaissance vehicles intended to supplant the Jeep in these roles (see chapter 6). Replacing this mixture of vehicles with M113s as an adaptation to Vietnamese mobility conditions, the regiment dealt with the problem of the vehicle's exposed and limited armament first by adding armor-plate shielding to the commander's .50 caliber machine gun, then adding two M60 7.62mm machine guns with their own shields, pintle-mounted one to each side of the rooftop cargo hatch.[37] Laden with extra ammunition and supplementary armor, notably a titanium antimine plate, these M113 ACAVs (Armored Cavalry Assault Vehicles) did yeoman service, acting as light tanks almost in the interwar sense of the term: highly mobile infantry surrogates carrying machine guns forward behind armor to engage dismounted forces and fix them for heavier weapons to attack.

The M113, whether employed as ACAV, APC, or support vehicle, was generally popular in Vietnam. One study by Lieutenant Colonel Roy F. Sullivan questioning "approximately 400 Armor and Infantry officers attending the 1968–69 regular course of the U.S. Army Command and General Staff College" and culled to "reflect only the answers given by Armor and Infantry officers who acknowledged having 'first-hand combat experience with the M113' "[38] is especially instructive. On the whole, the M113 was popular with all users. Averaging responses with respect to the vehicle's overall design—silhouette, frontal armor slope, height, and dependability—36 percent of armor officers polled rated it excellent, 48 percent satisfactory, and 16 percent unsatisfactory. Infantry respondents returned rankings of 20, 67, and 13 percent, respectively. It must be pointed out that the inclusion of the "dependability" parameter may skew the results positively. Armor respondents reported dependability at 74 percent excellent, 26 percent satisfactory, and 0 percent unsatisfactory. Infantry officers, in contrast, put dependability at 37, 60, and 3 percent, respectively. Whether the discrepancy involves differing expectations or differing maintenence standards between the branches is unknown.

Similarly, the M113's performance was also rated highly. Averaging among seven "capability" parameters—speed, cross-country mobility, cruising range, amphibious capability, air transportability, carry capacity, and ease of maintenance—survey results showed general approval of the M113: 50 percent of armor officers rated the machine excellent, 46 percent satisfactory, and 4 percent unsatisfactory. Infantry officers rated it 34, 58, and 8 percent, respectively. Armor officers, again perhaps more thoroughly attuned to dealing with the vagaries of tanks, skewed the maintenance figures: 58 percent rated the M113 excellent.[39]

This did not mean that no room was seen for improvement. In a second section of the survey, the respondents were heavily in favor of enhancing the M113's firepower and protection. Some two-thirds of the responents favored enhanced protection, but there was very little consensus on the priority of locations to be up-armored, aside from a decided preference on the part of armor officers for improved belly armor to protect against mines.

Similarly, half the infantry officers and seven-tenths of the armor officers wanted greater firepower, but there was scant agreement on the nature of the firepower enhancements desired. Respondents split between a 20–25mm cannon—capable of defeating field fortifications, suppressing heavy infantry weapons, and, in a different theater with more developed threats, as in the NATO center, engaging light armor and helicopters—and an antipersonnel area weapon such as a 40mm automatic grenade launcher or 60mm mortar. There was a bias toward the latter, especially on the part of the armor officers, who usually had less access to organic mortars than infantrymen. Only 9 percent of infantrymen and no armor officers (who, again, were accustomed to working closely in teams with tanks and their heavy guns) wanted an antitank missile; whether this related to Vietnam experience or the probable employment of the M113 in other theaters was not indicated. Most importantly, perhaps, about three-fifths of both armor and infantry officers polled felt that "see and shoot" capabilities (that is, facilities for the squad to observe and fire from under armor) would be a useful addition.[40]

In short, the officers were satisfied with the M113 as a vehicle, but felt it needed improvement as a fighting machine. The M113, of course, had never been intended as a fighting machine to begin with. Rather, it had been intended, by doctrine, to carry infantrymen forward strictly for dismounted combat; its successful use otherwise is powerful testimony to the flexibility of its design. This, however, was never touched on in Sullivan's discussion of his survey. In action, a fighting vehicle was what it had indeed become, regardless of doctrine or design. Its users, having used it in combat, saw a need for better capability in that area. To a considerable extent, combat experience suggested that the purist conception of the armored personnel carrier solely as a means of transport was a dead letter.

These veteran respondents were not alone in these ideas. As early as 1966, tests along these lines were being conducted at the Infantry School at Fort Benning, Georgia. An M113 was fitted with vision blocks and firing ports so that a rifle squad, seated back-to-back instead of face-to-face as usual, could use their M14 rifles and M79 grenade launchers while mounted.[41] These tests led to a more elaborate modification of the M113. This vehicle, the XM734, maintained these firing arrangements and added a turret mounting twin M73 7.62mm machine guns[42] in place of the open cupola-mounted .50 caliber weapon at the commander's station. This version was sent for trials in Vietnam, where it gained considerable user acceptance.

Field testing aside, however, the XM734 was strictly a testbed for the idea of mounted infantry combat. Four generations of armored personnel carrier design had made the M113 from which it was derived a protean and refined vehicle for the transport and support of infantry fighting on their feet. However, the modest level of protection and the boxy silhouette inherited from its progenitor was seen to make the XM734 unfit for the more intense armored battlefield envisioned. In the face of this threat, a different sort of vehicle had evolved, leading to the question of whether different skills might be needed to man it and different organizations to absorb it.

To summarize, the United States Army made considerable efforts to develop a specialized armored personnel carrier (APC), developing three generations of such machines during the 1950s. They were developed within a doctrinal framework that called for APCs to act as transports rather than tactical vehicles. Actual combat experience in Vietnam showed this doctrine to be overly restrictive; however, the current U.S. APC, the M113, proved fairly adaptable to mounted combat.

NOTES

1. Richard E. Simpkin, *Mechanized Infantry* (Oxford: Brassey's, 1980), 57.

2. "U.S. Infantry: Organization and Evolution," *Infantry* 66 (July–August 1976): 21.

3. House, *Toward Combined Arms Warfare*, 146–49.

4. Kaufman, "Organizations, Technology, and Weapons Acquisition," 121f.

5. R. W. Leonard, "The 'Queen' Makes Tracks," *Armor* 63 (September–October 1954): 9.

6. *Ibid.*, 10. The size of these doors was such that it was notably difficult for fully laden infantrymen to achieve this norm (Leland G. Andrews, "Chariot for the Queen," *Infantry School Quarterly* 46 [October 1956]: 82).

7. Bruce C. Clarke, "When the Armored Personnel Carrier Was Saved for the Army" *Armor* 80 (September–October 1971): 26.

8. David Pearson, "Ground Mobility," *Infantry* 62 (March–April 1972): 24.

9. "U.S. Infantry," 23.

10. Walter Millis, "Trojan Horse on Tracks," *New York Herald–Tribune*, 2 February 1954; reprinted in *Armor* 63 (March–April 1954): 28f.

11. Only a comparatively small number—1,780—of M75s had been built, first by International Harvester, then by the FMC Corporation. Clifford D. Bradley, "The Future IFV," *Infantry* 71 (July–August 1981): 23.

12. Pearson, "Ground Mobility," 24.

13. Bergen B. Hovell "Armored Infantry Is Different," *Army* 9 (November 1958): 49.

14. *Ibid.*, 53.

15. Leonard, "The 'Queen' Makes Tracks," 10.

16. *FM 17-20, Armored Infantry Units, Platoon, Company, and Batallion.* (Washington, DC: Headquarters, U.S. Department of the Army, 1957), 56–58, quoted in Ney, 46f.

17. Leonard, "The 'Queen' Makes Tracks," 11.

18. Clifford D. Bradley, "The Future IFV," *Infantry* 71 (July–August 1981): 21–27.

19. Leo A. Appling, Jr., "APCs of the Past" (letter), 72 (January–February 1982): 49f.

20. Kaufman, 139f. With the reorganization of the Army's technical services in August 1962 OTAC became the Army Tank-Automotive Command, or TACOM, the acronym generally used in this work.

21. Bradley, "Future IFV," 24.

22. *Ibid.*

23. William Boly, "The $13 Billion Dud" *California* (February 1983), reprinted in Dina Rasor, ed., *More Bucks, Less Bang: How the Pentagon Buys Ineffective Weapons* (Washington, DC: Fund for Constitutional Government, 1983), 25.

24. Kaufman, 141. Ordnance favored aluminum, CONARC steel.

25. Richard M. Ogorkiewicz, *Technology of Tanks* (Coulsdon, Surrey, England: Jane's, 1991), 367.

26. Pearson, 25.

27. Comparison of the air assault, or airmobile, infantry idea with the mechanized infantry idea is an interesting idea for further study. Unfortunately, it lies outside the

scope of the present work, which is occupied with establishing the dimensions and chronology of the mechanized infantry idea alone. A fair amount of work exisits on airmobility, most recently including Matthew Allen, *Military Helicopter Doctrines of the Major Powers, 1945–1992: Making Decisions about Air-Land Warfare* (Westport, CT: Greenwood Press, 1993); and Frederic A. Bergerson, *The Army Gets an Air Force* (Baltimore: Johns Hopkins, 1980). In particular, the contrast between the airmobile attitude toward its transport, an independent organizational element, and the mechanized infantry's of the transport, as a piece of equipment organic to the lowest organizational level, warrants examination.

28. "Reorganization of Army Divisions," in USACGSC, *Development of Combat Divisions*, v. 2:1-71f.

29. Pearson, "Ground Mobility," 23.

30. Later doctrinal publications were under the Infantry 7-series, but still separate; FM 7-17 for the armored infantry platoon and squad, for instance. See Thomas G. Woods, "MOS 1560," *Armor* 64 (July–August 1955): 27. Captain Woods, a staff officer at the Armor Center, Ft. Knox, Kentucky, considered FM 7-17 inferior to the corresponding manual for tank platoons, FM 17-32.

31. This was, in fact, the title of two articles during the 1950s in official publications, one by Captain Charles W. Koburger, in *Armor* 60 (January–February 1951): 40–41, and one by Lieutenant Colonel Bergen B. Hovell in *Army* (November 1958): 49–53.

32. James T. Burke, "Armored Infantry and Recon Unit Organization," *Armor* 65 (July–August 1956): 16.

33. Donn A. Starry, *Mounted Combat in Vietnam* (Washington, DC: Department of the Army, 1978), 10.

34. *Ibid.*, 22f.

35. Steven J. Zaloga and James W. Loop, *Modern American Armour* (London: Arms and Armour, 1982), 52f.

36. Eric M. Bergerud, *Red Thunder, Tropic Lightning* (Boulder, CO: Westview Press, 1993), 68.

37. Starry, *Mounted Combat*, 73.

38. Roy F. Sullivan, "APC in RVN," *Armor* 80 (January–February 1970): 36.

39. *Ibid.*, 37.

40. *Ibid.*, 38.

41. Thomas P. Kehoe, "The M113 Armored Porcupine," *Infantry* 56 (January–February 1966): 24.

42. The M73 was the weapon used as the coaxial machine gun of the M60 tank, among others. Although specially designed for armored vehicle use, it was unpopular and ultimately unsucessful due to reliability problems.

Chapter 4

The NATO Center and Early Infantry Fighting Vehicle Development, 1960–1964

The search for new "concepts" has been underway for 20 years, and nothing of great consequence has appeared. This is not because there are no new concepts available. The real problem lies with those who are hunting for something new. They want a new look, but one that is familiar enough to suit their own tastes.

—Robert P. Ziegler[1]

By the early 1960s, the U.S. Army had spent two decades expecting to fight in Europe while having to equip itself to fight elsewhere. This situation did much to dictate the design of major equipment, as can be seen in the foregoing discussion of the evolution of the armored personnel carrier, especially the M113, with its design premium on airportability. Richard Simpkin summarized the U.S. strategic situation nicely in 1979:

Two factors combine to put United States thinking on the tank in a different ballpark than her European allies and potential opponents. For her the age of the tank has coincided with superpower status and the maritime and aerospace commitments that go with it. The land battle rates comparatively low in the USA's priorities, as it did in those of imperial Britain; and the tank does not necessarily rate first in her land armory. Now she is focusing on the European theater again. . . . Likewise the United States' worldwide scope imposes a number of special operational and logistic requirements on her tanks. Arctic and tropical capabilities are not just add-ons or sales gimmicks but genuine basic needs; the retention of serviceability over long periods in the hold or tank deck of a ship is another. Restrictions on loading gauge and weight which can be effectively ignored in Europe may be critical for sea or air transport and/or in less developed theaters. American tank design is subject to many compromises; it has to aim at a balanced and versatile vehicle.[2]

While Brigadier Simpkin speaks here of tank design, his point extends to armored vehicles in general, and infantry vehicles in particular.

The U.S. Army's land warfare doctrine of the late 1940s and 1950s can be characterized as a steady-state evolution from wartime experience. It envisioned a more mechanized version of the Army's performance in the European Theater of Operations, punctuated by the brief, nuclear-obsessed Pentomic era. This mindset can be seen in the major equipment of the Army's maneuver arms entering service at the time of the early 1960s ROAD initiative.[3] The M113 APC was functionally almost indistinguishable from its predecessor, and nearly so from the two before that, being designed as an armored transport rather than a fighting vehicle. Similarly, the M60 tank was in essence the fourth-generation version of a late–World War II medium tank design, albeit heavier and more powerfully armed. This relative doctrinal stasis was encouraged by economies imposed on conventional forces by the "New Look" strategic policies of the Eisenhower administration, which relied on nuclear deterrence. This situation became subject to change, however, as Kennedy-era strategic ideas of flexible response and the reassessments of defense policy associated with the Berlin crisis of 1961 served to focus new attention on the European theater and its peculiar conditions. Although worldwide contingency missions still received attention, the possibility of European operations within the context of the NATO alliance received new attention as the Army planned the acquisition of a new generation of equipment.

Arguably, attention was needed. NATO faced a challenging military situation. Aside from the threat—rather, the promise—of a nuclear, biological, and chemical (NBC) environment and the technological sophistication of the antagonists, the most salient features of the putative European battlefield were cramped space and an enemy endowed with considerable numerical superiority. Numbers were subject to change, but one West German commentator put the matter succinctly:

> The Federal Republic of Germany is in a precarious geostrategic position. The border region between Luebeck and Passau is roughly 1,000 miles long and contains long stretches of open country. It is not sufficiently deep from east to west for military operations, while 40 pecent of the territory, 30 percent of the population, and 24 percent of the West German industrial potential lie within 60 miles of the border.
>
> In a crisis, 24 divisions—12 allied and 12 *Bundeswehr*—equipped with 5,000 main battle tanks and supported by 2,200 firstline combat aircraft would be available for the defense of the Federal Republic. At most, six additional divisions would be available within two weeks through mobilization or airlift from the United States. Including these reinforcements, each division would have to defend a sector 60 to 70 kilometers wide against an enemy three times as strong.[4]

One consequence of this renewed appreciation of the conventional threat in Europe was the initiation of a collaborative tank-development project with the Federal Republic of Germany, officially agreed upon in August 1964. Although the two parties differed on many of the details, it was evident that the new tank was to be a radical departure from the M47/48/60 "Patton" family of tanks that had been the backbone of NATO's armor since the alliance's inception. The MBT-70 (Main Battle Tank-1970s) was to embody significant improvements in firepower, mobility, and protection, especially radiological protection. Its goal was to use advanced technology to redress the numerical imbalance. A new tank, especially a more mobile one, in turn, implied a requirement for a new vehicle to carry the infantry working with it. By December 1964, a long range intelligence forecast led to the promulgation of a new mechanized infantry doctrine to reflect the reappraisal of conditions in Europe.[5] The change of doctrine, combined with the prospective changes in equipment was problematic: the current doctrine, and the armored personnel carriers to support it, had been evolving over almost two decades—the length of many entire military careers. The new doctrine had further encompassed four generations of evolving equipment, culminating in the inexpensive, reliable, and wildly adaptable M113. Institutional movement from this comfortable position, however necessary, could hardly be seen to be a simple matter.

Certainly the West Germans, who were rapidly assuming enhanced importance in NATO in the 1960s, thought that new equipment new organization, and new—or rather, renovated—doctrine were all needed to meet the alliance's military situation. The army of the Federal Republic of Germany, the *Bundeswehr*, had, since its formation in 1956, resisted U.S. armored infantry concepts, with their purist emphasis on dismounted combat as the sole mode in which armored infantry was to engage the enemy. The West Germans had instead insisted that infantry working with tanks should fight mounted whenever possible to maintain the speed of armored forces. This belief was based on negative *Wehrmacht* experience in World War II, especially on the Eastern Front, where the shortage of specialized *Panzergrenadier* vehicles had revealed the problems of employing motorized but unarmored infantry in conjunction with tanks (see chapter 2). The Germans thus had little experience of the benefits of armored infantry per se. As Richard Simpkin wrote: "The Third Reich's army . . . bequeathed to the Bundeswehr the doctrine that it needs to be armored and capable of bringing fire to bear from under armor when mounted. But this seems to be a case of 'Do what I say, don't do what I do.' "[6]

On the strength of their experience with armored and unarmored *Panzergrenadiere* the *Bundeswehr* that replaced the *Wehrmacht* declined the offer of U.S. M59 and M113 armored personnel carriers early on; instead, one of the first projects of the new army was a new infantry vehicle for its *Panzergrenadier* forces. The *SPz* (*Schuetzenpanzer*) 12-3 was considered more than a means of transport for the *Panzergrenadier* squad. Rather, it was an integral, even defining piece of the squad's equipment, its members being expected to stay with it under normal conditions in both attack and defense.[7] Being intended to work closely with tanks, which were scarcely expected to float, it did without amphibious capability. It was, however, considerably more heavily armored than the M113, weighing 4 tons more despite its smaller personnel complement of a driver, gunner, and six dismountable troops. It also presented a smaller silhouette, two feet lower in height even including its turret-mounted 20mm automatic cannon. This weapon enabled the crew to engage light armored vehicles, antitank weapon emplacements, and low-performance aircraft such as helicopters. These were awkward targets for tanks: too small or nimble to engage efficiently with a 90 or 105mm main gun, yet too tough or distant for coaxial machine gun fire. The cannon was therefore a useful complement to the tank gun when the *SPz* was operating in close support of tanks, rather than a surrogate for a tank gun or an extrapolation from an infantry weapon.

Nor was the *SPz* reliant solely on its fixed armament; the squad members could fire their personal weapons, which included light machine guns, selective-fire G3 automatic rifles, and light antitank weapons, as well. To do so, however, they had to open the roof hatches of the troop compartment, exposing themselves to hostile action and sacrificing overhead cover. This was not considered a great drawback in conventional conditions; it was expected that the danger of artillery fire could be discounted if the enemy was close enough to be engaged with small arms.[8] Firing from open hatches was, however, a potential problem in an anticipated NBC environment: it would expose the interior of the vehicle to contamination by chemical agents and radioactive fallout, thus compromising the protection afforded by the vehicle.

It is worth noting at this point that the *Bundeswehr* did not regard the *SPz* as a universal infantry vehicle, recognizing the limitations as well as the capabilities of the small, heavily armed squads it fielded. Even in fully equipped *Panzergrenadier* brigades, which consisted of three infantry, one tank, and one artillery battalions, only two of the three infantry units were equipped with the *SPz* 12-3.[9] The third was a motorized battalion fighting strictly on foot and designated *Panzergrenadiere (mot)*, as opposed to the armored *PzGren* (*SPz*), that is, motorized rather than armored infantry. This motorized unit was first tranported in Mercedes *Unimog* trucks. Later it gained M113s, and the new equipment brought a new designation: *Panzergrenadierbattaillon (MTW)*, "MTW" (*Mannschaftstransportwagen*) being the German term for tracked armored personnel carriers. The battalion thus gained much of the tactical mobility of the other *Panzer* forces, plus a limited ability to cross water obstacles. One *Bundeswehr* officer described the logic of the mixed force thusly:

> Oversimplified, this means: with the exception of the two battalions of Panzergrenadiere (mot), all Kampftruppen[10] of a division and most of the division's artillery, antiaircraft, engineers, etc., are fully tracked and in a position to turn defense into several local attacks. In other words, while, for a given reason, a division may be engaged in defense, its brigades may lead a mobile fight, the better portion of which results in attacks with limited objectives. The positive effects of this concept on the morale of combat troops are obvious.
>
> The purpose of this deliberately pointed interpretation of current West German Army doctrine is not to be construed as a devaluation of the Panzergrenadiere (mot). The

motorized Infantry of the Bundeswehr is undoubtedly here to stay. Who else can better fight in defense positions and against dug-in enemy Infantry at close range? Who else is better trained to fight in close combat, particularly at night or in poor visibility, attacking or defending? There is a need for this type of Infantry—always and everywhere. In most countries, it is the only type of Infantry available.[11]

One such country was the United States. American armored infantry vehicles, due to the doctrinal insistance that infantrymen were only to fight dismounted, were less suited to mounted combat than those of any other armor-producing nation save Great Britain. As a result, whatever the drawbacks of the *SPz*, it attracted the attention of American observers, many of whom saw even the M113 as less than optimal for the highly fluid type of armored warfare forecast. Throughout the 1950s, with the M75 and M59, there had been debate in armored infantry circles over when to dismount the infantry; the *Bundeswehr* was asking whether to dismount them at all.

The new idea reached the Army professional press with the February 1963 issue of the U.S. Army Command and General Staff College's *Military Review*, in which Captain Clinton Granger issued a manifesto of sorts. His article: "Wanted: An Infantry Fighting Vehicle" recounted the history of U.S. infantry vehicles to date, pointing up the lack of equipment to support mounted combat. Doctrine, Granger observed, had in essence adapted itself to existing technology, as this lack was the result of the institutionalization of a design flaw. The lack of overhead hatches, firing ports, or vision devices in the M44 and its successors had precluded the possibility for mounted combat by armored infantrymen. Granger asked: "The basic question remains: Is an armored personnel carrier a species of truck, or is it a more flexible item, with capabilities as a tactical combat vehicle within the well-recognized limitations of armor and armament?"[12]

Citing the West German *Schuetzenpanzer* as an example, and citing Rommel's and Patton's ideas about the value of fire in meeting engagements, Granger then suggested that the M113, then just entering service, could be minimally modified to allow the mechanized infantrymen inside to fire from under armor:

Since the carrier[13] serves as a tactical vehicle in a river crossing, why not exploit its potential elsewhere too? To expand the role of the carrier the first step must be to add the two characteristics noted earlier: a design which enables passengers to see what is going on around the carrier, and to use their individual weapons. The other aspects of recent carrier development are perfectly sound—regardless of what specie [*sic*] you may consider the carrier to be [i.e., transport or fighting vehicle, referring to improved layout and amphibious capability]—and the addition of the capabilities noted will not detract from the use of the carrier in other than the infantry fighting vehicle role.[14]

Granger did not treat the probable need for increased protection for the infantry vehicle, and in fact argued against excess sophistication—"complex electronic viewing equipment, TV screens, or individual turrets for every rifleman"[15]—as both unrealistic and unnecessary. In effect, Granger thus anticipated the XM734 (see chapter 3) several years beforehand: a minimally modified M113, rather than a purpose-built *SPz*-type vehicle.

The Army's development apparatus, however, was thinking more expansively. As early as 1958, the U.S. Army Infantry School had recommended the development of an "infantry fighting vehicle," or IFV. The Infantry School's concept was more elaborate than either the West German *Schuetzenpanzer* or the M113 derivative suggested by Captain Granger in *Military Review*, combining features of both machines. The infantry fighting vehicle was envisaged as carrying a 20mm cannon and a 7.62mm machine gun, as well as being fitted with firing ports for the individual weapons of squad members. The Infantry School also advocated an antitank capability for the vehicle, whether stowed as a squad weapon or fitted to the vehicle itself.[16]

The Infantry School's 1958 study bore no fruit directly, but by 1964 the U.S. Army and the *Bundeswehr* had carried out collaborative studies of mechanized infantry employment. These works endorsed mounted combat by mechanized infantry under appropriate conditions, a change in doctrine soon put forward in publications such as *FM 100-1*, which set out the functions of the various component arms and branches of the service, and the Infantry mission statement in Army Regulation 10-5. The same year, experiments at the Combat Development Command Experimental Center at Fort Ord, California, concluded that effective fire could be delivered from a moving vehicle.[17] This doctrinal shift suggests that even with projected improvements in mobility and protection, the M113 would likely be tactically incompatible with the MBT-70 main battle tank, that is, if the projected vehicle itself was not evidence enough.

By 1964, the new tank's design was taking its final shape. The chosen concept had been one of the less radical of the proposals. It retained the relatively traditional configuration of a rear-engined hull and manned turret carrying the armament. Resemblance to earlier tanks ended there, however. The hull was to contain little more than the drive train and engine—a novel variable-compression-ratio diesel—all riding on an equally novel hydropneumatic suspension that allowed the tank to "crouch" and lower its silhouette. To simplify the provision of NBC protection, the hull did not contain a driver, who instead drove the tank from a counterrotating capsule within the turret; the elaborate mechanical arrangements necessary to accomplish this tended to negate the gains in simplicity made by consolidating the crew. To accommodate the driver, in turn, the turret contained no crewman to act as loader, the gun being served by an automatic loader in the turret bustle. The gun, yet in its turn, was no ordinary gun, but a 152mm hybrid weapon firing both conventional ammunition—"conventional" encompassing antipersonnel flechette and armor-piercing fin-stabilized discarding sabot rounds—and Shillelagh antitank guided missiles. Given that a 152mm gun was likely to be an overmatch for many anticipated targets, the traditional .50 caliber machine gun at the commander's station was to be replaced by a retractable, remote-controlled 20mm cannon (the same model as that on the German *SPz*) firing a choice of armor-piercing and high explosive ammunition.

Even before the MBT-70 concept was defined, it appeared that a new sort of mechanized infantry vehicle, one permitting mounted combat, would be useful on the European battlefield foreseen; the prospect of a tank designed for this environment appeared to make one imperative if the latter was to be given adequate support. In January 1964, in recognition of these changes in doctrine and equipment, the Office of the Chief of Research and Development of the Department of the Army directing that development of a new Mechanized Infantry Combat Vehicle be commenced.[18] To this end, studies were initiated for a mechanized infantry vehicle to complement the MBT-70, which would include NBC protection and a capability for crew and passengers to fight from under armor. Mobility was to be comparable to that of MBT-70; in addition, the new vehicle was also to to retain the amphibious capability of the M59 and M113, in contrast to the *Bundeswehr*'s rationale for dispensing with this capability. Similarly, the squad it carried was to be the standard infantry strength of eleven, rather than the smaller squad the West Germans favored for their *Panzergrenadiere*.

Development was to be carried out in two stages: an interim "Mechanized Infantry Combat Vehicle" (MICV-65) was to prove the concept for a full-up production vehicle, MICV-70. A contract was let for MICV-65 prototypes to Pacific Car and Foundry Company of Renton, Washington, in the spring of 1964.[19]

The MICV-65 testbed vehicles, designated XM701, bore a vague resemblance to their armored personnel carrier predecessors, the M59 and M113, with a bluff bow and ramped stern. Mechanically, though, they derived from the M110 8-inch self-propelled howitzer, using its power train, suspension, and tracks along with the final drives[20] of the M109 155mm self-propelled howitzer; significantly, the Army Materiel Command office responsible for developing MICV-65—Program Manager, Combat Vehicles—also held responsibility for self-propelled artillery. There were other differences as well. The XM701 was much heavier than the M113, weighing 49,000 pounds in aluminum and 52,000 pounds in steel compared to the earlier vehicle's 24,000 pounds. It was also notably larger and more heavily protected. The XM701's front was designed to resist 14.5mm heavy machine gun fire all around, as well as 155mm high-explosive fragments from the top, and 12.6-pound mines from underneath.[21] This was a significant improvement over the 7.62mm level of protection designed into the M113. The greatest departure, though, was in armament. Amidships, the XM701 carried a two-man turret armed with a 20mm cannon, together with a 7.62mm coaxial machine gun. This mounting could accommodate a stabilization system to allow the armament to be fired effectively on the move. The nine men in the passenger compartment—sloped at the sides for better ballistic protection, in contrast with the slab-sided armored personnel carriers—could use their personal weapons from under armor as well. They fired through seven hull firing ports,[22] three on each side and one to the rear, aiming through vision blocks provided for the purpose.

In engineering and service tests conducted from July through December 1965,[23] much valuable experience was gained with the XM701, but the Army declined to order it into production, even on the interim basis originally intended.[24] Tests showed inadequate tactical mobility (especially cross-country speed) relative to the MBT-70, despite the XM701's 400-mile range and the ability to clear a 36-inch vertical obstacle. In addition, the XM701 was considered to have inadequate strategic mobility: 124 inches wide, it could not fit into the C-141 transport aircraft[25] It might have been argued that neither would the MBT-70 it was supposed to accompany into battle, but the point was moot in the face of official requirements. The XM701 test vehicles were thus consigned to the role of equipment testbeds for the evaluation of armament and propulsion components.

The MICV-65 prototypes had been found unsatisfactory. The concept of the mechanized infantry combat vehicle itself, though, was nonetheless still thought to be sound. A new round of studies for an improved vehicle, MICV-70, was begun in July 1965. Tentative operational characteristics were assigned the following month; by September, the projected MICV-70 gained the official designation "Combat Vehicle, Mechanized Infantry, XM723."[26]

U.S. force planners, then, were worried about the applicability of a doctrine restricting infantry to dismounted combat in the face of an NBC battlefield environment. Encouraged by a similar development program by the West German *Bundeswehr*, a program to develop the new vehicle type, called a mechanized infantry combat vehicle (MICV) was launched. The first attempts to produce such a vehicle were unsatisfactory, as the developers insisted on greater capabilities at low prices without concomitant design tradeoffs being made; however, the development process continued. It did so slowly, since the growing U.S. involvement in Vietnam starved the program of funds. Once again, however, foreign events bore heavily on the direction of the program.

NOTES

1. "Mechanized Fighting Vehicle," *Military Review* 46 (July 1966): 3. Colonel Ziegler was chief of the Organization and Doctrine Division, Headquarters Combined Arms group, Fort Leavenworth, Kansas at the time.

2. Richard E. Simpkin, *Tank Warfare: An Analysis of Soviet and NATO Tank Philosophy* (London: Brassey's, 1979), 63.

3. See chapter 2.

4. F. O. Miksche, "Effective Antitank Defense" *Military Review* 51 (December 1971): 48.

5. "Historical Milestones " (unpaginated memo, hereafter given as "Milestones") component of: United States. Department of the Army. Project Manager, Mechanized

Infantry Combat Vehicles. "Project Master Plan: Mechanized Infantry Combat Vehicle, MICV—XM723." National Archives and Records Administration, Federal Records Center, Suitland, MD, Box 338-90-512, "PM-FVS PL & ANAL 1969." Hereafter "Master Plan."

6. Richard E. Simpkin, *Mechanized Infantry* (London: Brassey's, 1980), 24.

7. H. J. Thiele, "*Panzergrenadiere*," *Infantry* 53 (November–December 1963): 42.

8. *Ibid.*

9. *Ibid.*, 39f. This organizational principle held at higher levels as well; the normal *Bundeswehr* division under this structure, with the exception of light infantry mountain troops, was built around two such *Panzergrenadierbrigade* and a *Panzerbrigade* of two tank and one *SPz*-equipped *Panzergrenadier (PzGren)* battalions.

10. Tank and infantry battalions; the *Bundeswehr* made no institutional distinction between armor and infantry as separate combat arms.

11. Thiele, "*Panzergrenadiere*," *Infantry* 53 (November–December 1963): 40.

12. Clinton E. Granger Jr., "Wanted: An Infantry Fighting Vehicle," *Military Review* (February 1963): 32f.

13. M113.

14. Granger , 32.

15. *Ibid.*, 34f.

16. United States. Army. Infantry School. Combat Developments Office. *Infantry Fighting Vehicles for the PENTANA Combat Group* (Ft. Benning, GA: U.S. Army Infantry School, 1958), 3–4; cited in Daniel Joseph Kaufman, "Organizations, Technology, and Weapons Acquisition: The Development of the Infantry Fighting Vehicle" (unpublished diss., Massachusetts Institute of Technology, 1983), 165f. Kaufman considers the requirement for antitank capability "prophetic" in view of later events (see chapter 7). This can, however, also be seen as a continuation of equipment practices for armored infantry squads going back as far as World War II; see chapter 4 for details. More prophetic is the study's suggestion that the desired characteristics be achieved within a limited weight through use of a reduced squad. The proposal foundered on the Armor School's insistance on a full–size squad. This turn of events was a curiosity in view of Armor's history of trading manpower for firepower and mobility, and an anomaly in view of Infantry's preference for standardized organizations (see chapter 3). Kaufman, plausibly, suggests this attitude arose from a parochial desire on the part of the Armor branch to continue its monopoly of armored combat vehicles.

17. Jeff F. Cherry, "Mounted Combat," *Infantry* 65 (September–October 1975): 13.

18. United States. Army. Tank-Automotive Command. *U.S. Army Tank-Automotive Command Annual Historical Review, 1 October 1981–30 September 1982*. Joseph L. Avesian, Joseph Mihalak, Ann M. Bos, comps. Warren, MI: United States Army Tank-Automotive Command, 1983, 219; hereafter given as "TACOM AHR'83."

19. Bradley, "Future IFV," 25. Designated XM701, a prototype of the MICV–65 was delivered in May of 1965. It was followed by five pilot vehicles—two aluminum-hulled and four steel in all (*AFJ*, 5 June 1965, 22).

20 . That is, the components delivering power from the transmission to the tracks.

21. U.S. Army Weapons Command, *Technical Development Plan, Mechanized Infantry Combat Vehicle–65* (Rock Island, IL: U.S. Army Weapons Command, 1966), 6.

22. Sullivan, "Muddling," 9.

23. TACOM AHR'83, 223.

24. *Technical Development Plan, MICV–65*, 7f.
25. Sullivan, "Muddling Through," 9.
26. "Milestones."

Chapter 5

The Warsaw Pact Threat and U.S. Infantry Fighting Vehicle Development, 1967–1975

> Studies which originate in and serve a particular service tend to a narrowness of focus and an orthodoxy of assumption that is both understandable and conceptually paralyzing.
>
> —Colin S. Gray[1]

> The irregular verb "to design"
>
> I create
> You interfere
> He gets in the way
>
> We co-operate
> You obstruct
> They conspire
>
> —David K. Brown[2]

By November 1967, the Army had been developing mechanized infantry combat vehicles for nearly a decade without being close to fielding one. In that month, however, it became evident that the United States Army was not the only one envisioning a high-intensity, mechanized, and possibly nuclear battlefield and preparing to fight on it, as the Soviet Army's BMP (*Boevaya Mashina Piekhota*, infantry combat vehicle) paraded in Red Square for the first time.

Despite a series of tank designs going back to the 1920s, the Soviet Army had come relatively late to other aspects of mechanization. Its Ground Forces had not abolished horse transport until 1955. Within two years, however, the Army had committed itself to the abolition of regular infantry, declaring all infantry save paratroops to be "motor-rifle" formations. These troops were to be carried by their own organic armored transport. The initiative was nominally completed by 1963, at least for first-line formations.[3] This ambitious plan, however, was implemented with a lackluster assortment of equipment. The infantry components of tank divisions and regiments used the BTR-50 (*Bronetransportr*, armored transport), a derivative of the PT-76 amphibious light tank. The mainstay of the motor rifle divisions, where they were available, were mainly armored trucks, such as the BTR-40 and -152. It was 1961 before a purpose-built APC, the eight-wheeled BTR-60, reached the motor rifle divisions. All these vehicles began production as open-topped designs carrying half-platoons, which mounted and dismounted over the sides. As greater numbers became available, however, later models of both the BTR-50 and the BTR-60 gained top armor, at a concomitant cost in troop capacity, the former half-platoon vehicles becoming squad vehicles. In addition, the BTR-60, in its definitive version, the BTR-60PB, gained a turret manned by a dedicated gunner armed with one 14.5mm and one coaxial 7.62mm machine gun. In some models, firing ports for the embarked motor rifle squad pierced the sides for additional combat capability.

Overhead cover gained special prominence as Soviet doctrine came to embrace tactical nuclear weapons, the so-called "revolution in military affairs" of the early 1960s. The need to conduct armored offensive operations supported by the extensive use of nuclear and chemical weapons and conducted in the face of them as well demanded an infantry vehicle with characteristics different from those of the current inventory. In particular, such a vehicle was seen to require high speed, enhanced cross-country mobility, and the ability to engage in mounted combat, all so that it could conduct missions with minimal exposure of the squad to the hostile NBC environment. The BMP was designed and fielded to fulfill this requirement. As it happened, the height of the Soviet enthusiasm for tactical nuclear weapons peaked under Nikita Khrushchev, and had begun to wane by 1967;[4] nonetheless, the vehicle retained the characteristics. The BMP was based on the same power train and running gear as that of the earlier BTR-50 and that of the PT-76 amphibious light tank, albeit now in a front-engine configuration. Because of its lighter weight and smaller size, however, it improved somewhat on the earlier vehicles' automotive performance. Additionally, it retained its predecessors' amphibious capability, and provided collective NBC protection for its occupants—a commander, driver, gunner, and eight infantrymen.

It was in the area of firepower, however, that the BMP represented the greatest break with past practice. It was fitted with ten NBC-resistant firing ports, four to each side and two to the rear, each with its own periscope. From these, the squad could employ its light machine guns and Kalashnikov assault rifles, as well as from four two-man roof hatches suitable for the use of rocket-propelled grenades and other weapons with backblasts. This capability, however, was dwarfed by the weapons fitted to the one-man turret. This carried a 73mm smoothbore gun (erroneously first reported in the West as the 76mm tank cannon of its progenitor, the PT-76 light tank) firing the same hollow-charge, rocket-assisted antitank projectiles as the standard infantry recoilless antitank gun, a rail for launching Malyutka ("AT-3," in U.S. intelligence parlance; "Sagger" to NATO) antitank guided missiles, and a coaxial 7.62 x 54mm machine gun. A platoon of three BMPs and crew thus fielded considerably more antitank firepower than a motor rifle company equipped with BTR-60s. It was, in effect, quite literally an "infantry fighting vehicle," conveying an essentially standard infantry squad under armor, along with an enhanced complement of essentially standard company- and platoon-level heavy weapons: light antitank guns and missiles and medium machine guns.

All this capability, of course, came at a considerable price, both in monetary outlay and in design trade-offs. Armor protection was but little advanced over the BTR-50, although the BMP presented a smaller target. Small size, though, meant that the vehicle crew and infantry squad were packed tightly together, to the point that the motor riflemen could be no more than 1.7 meters tall.[5] Ammunition and fuel were packed in tightly as well, to the extent that the dual rear doors doubled as fuel tanks.[6] As a result, the BMP presented virtually no aspect not vulnerable to enemy fire; practically any penetration of the armored volume would endanger crew, fuel, or ammunition. Packing all this capability into such a small envelope also inevitably meant that the vehicle's systems quite literally got in each other's way. The BMP commander, who doubled as the motor rifle squad leader, was seated behind the driver; thus, the turret blocked his view over a wide sector to the right rear, in contrast to the all-around view afforded the commander of a more conventionally laid-out vehicle. At the same time, though, over his own best arc of observation, no turret weapons could be brought to bear due to the protrusion of the commander's infrared searchlight, periscope, hatch, and occasionally his head as well as that of the driver. The separation of the commander and gunner made fire control difficult, and when the commander dismounted with the squad, as prescribed by doctrine, the gunner also had to command the vehicle, imposing a substantial workload under combat conditions. The profusion of weapons, of course, also represented a grave training problem for a mass conscript army based on two year terms of service and six month cycles of instruction.

Planners and developers in the United States, however, were more concerned with the potential capabilities of the BMP than its drawbacks. Given the controlled exposure to the new weapon under Cold War propaganda conditions, the former were far more evident than the latter at the time. The specter of being outclassed both quantitatively and qualitatively on the projected European battlefield led to an expansion of the Mechanized Infantry Combat Vehicle (MICV) program.

The MICV program had continued after the initial testing of the rejected XM701 pilots, but at a low level. The pilot vehicles had continued in service as testbeds for subsystems: auxiliary waterborne propulsion devices such as waterjets and Kort ducted propellers, weapons stabilization systems, and hydrodynamic improvements to reduce the bluff-fronted vehicle's bow wave and thus enhance both the driver's vision and the usability of the main gun while afloat.[7]

At the same time, development of the MICV-70 went ahead on paper. In September 1965, Cornell Aeronautical Laboratory commenced a Parametric Design Cost Evaluation study[8] of fifty-three vehicle concepts submitted by industry and government.[9] Even this large number represented a serious winnowing; in the words of TACOM design engineer Clifford Bradley, between 1963 and 1972, the TACOM Advanced Concept Group alone

> . . . generated approximately 145 vehicle system concepts with supporting analytical work . . . including concepts for vehicles as small as 16,000 pounds and as heavy as 100,000 pounds for those with NBC protection. It explored half-squad vehicles and two-squad vehicles; front power plants, rear power plants, centrally mounted power plants, and split power plants: engines rear, transmissions in front, and just the opposite. It explored various levels of armor protection, with automatic components to match each weight resulting from the selected armor level.[10]

At the same time, however, the main thrust was fairly clear from the Army Weapons Command's Technical Development Plan. This document, issued in February 1966, called for an amphibious squad carrier with cross-country mobility to match the MBT-70[11]—in short, a next-generation analog of the M113, intended to replace the older vehicle one-for-one.[12] At the same time, however, the Army's equipment developers were seeking to fit the capabilities sought for the MICV-65 into a more acceptable package than the XM701. In this, they were supported by the Cornell Parametric Development/Cost Effectiveness Study, the results of which pointed to a 12-man aluminum-armored vehicle with a 20mm cannon as the most feasible design choice for the MICV-70.[13]

The bulk of the design studies for MICV-70 were concluded in February 1968.[14] Regardless of their extent, there was still considerable disagreement over the characteristics of the proposed vehicle. The Army research and development establishment, in particular the Combat Development Command (CDC), favored the fairly ambitious vehicle set forth in the original MICV-70 Quantitative Material Requirement (QMR). This was a conceptual design featuring protection against 23mm gunfire, stabilized armament, and collective NBC protection. The Army operating agencies, in particular CONARC, were concerned by the bulk and complexity of such a vehicle, and supported a more austere approach to the MICV. CONARC was supported in this position by the Deputy Chief of Staff for Logistics and the Comptroller of the Army. These disagreements culminated in the disapproval of the QMR by General Ralph Haines, the Army's Vice Chief of Staff.[15]

The disapproval of the QMR did not eliminate the perception on the part of the parties involved that an M113 successor was still needed. To resolve these conflicts, in August, 1968, the Army Chief of Staff had a Mechanized Infantry

Combat Vehicle Ad Hoc Study Group set up. This body (generally referred to as the "Casey Board" after its director Brigadier General designate George Casey) was to investigate the MICV concept, both as to whether the requirements involved could be carried out by a single vehicle type, and whether the weight, size, and complexity of the vehicle—and therefore its cost—could be reduced without unacceptable losses of capability. Significantly, approved mechanized infantry doctrine was not to be challenged by the board.[16]

The Casey Board reported within a month. Largely because of the prohibition on challenging doctrine and emphasis on simplicity in the board's charge, its conclusions were straightforward and predictable. In essence, they supported the Cornell studies and the MICV-70 Technical Development Plan as before. The board reported that the MICV requirement could be filled by a single vehicle. This vehicle, in turn, could form the basis of a family of ancillary vehicles, analogous to the numerous variants of the M113. Ironically, in view of later developments, further product improvement of the M113 was recommended against due to the logistical impact of maintaining two vehicle fleets. The Casey Board did suggest some austerities. Reductions in battlefield endurance from 48 hours without resupply to 24 were proposed, along with modest sacrifices of automotive performance and armor protection. These led to a smaller and cheaper vehicle: 36,350 pounds and $108,900 for the Casey board MICV versus 44,000 pounds and $149,500 for the machine as originally planned.[17]

All this deliberation and the redesign it engendered took time. The appearance of the BMP seemed to call for faster action, at least to produce an interim vehicle. Army planners were thus compelled to examine the usual two options normally available to procurement institutions in a hurry: either purchase foreign equipment or adapt that on hand. The first of these options was not terribly attractive, as only one Western vehicle was available; this was the West German *Schuetzenpanzer-Neu*, later designated *Marder* (Marten).

Just entering preproduction, the *Spz-Neu* had been in development since 1960 as a successor to the *Spz 12-3*, attempting to answer many of the earlier vehicle's perceived deficiencies in mobility and protection. It thus incorporated many of the design objectives of the MICV-65. It boasted enhanced protection for the *Panzergrenadier* squad, including comprehensive defenses against NBC threats, together with ballistic protection approaching that of the Leopard tanks it was to work with, although to be sure the Leopard made considerable sacrifices in this respect to keep weight down. *Marder*'s mobility, too, was comparable to that of the Leopard. The new vehicle carried the 20mm cannon of its predecessor, now mounted in an innovative two-man turret with the gun and a coaxial machine gun carried overhead rather than in front of the turret crew. In addition, it gained the long-sought capability for carried troops to fight from under armor. Four firing ports, two to each side with six associated overhead periscopes, allowed 9mm Uzi submachine guns to be used for close combat, their shorter length making them preferable to the G3 automatic rifles used when dismounted. These weapons were supplemented by a second 7.62mm machine gun mounted on the rear roof and fired from inside the vehicle.

Unfortunately, from the American perspective, the *Marder* also embodied many of the drawbacks of the XM701. Because the *Bundeswehr* faced none of the considerations of strategic mobility so important to the U.S. force, the vehicle weighed 31 tons (combat weight) to the XM701's 26, was 128 inches wide to the U.S. vehicle's damning 124, and lacked the required amphibious capability. For all that, the vehicle dismounted six men with a crew of four—a commander,a gunner, a driver, and a second gunner operating the rear machine gun—as opposed to the nine and three projected for MICV-70. *Marder* was therefore a complex and bulky vehicle by any standard; in effect, an example of the computer scientist's "second-system effect," where the problems of accommodating cumulative minor improvements added to a new design eventually obscure the worthwhile features of its predecessor. It was certainly a complex and bulky one by the relatively austere standards of the U.S. Army and its congressional paymasters. In the words of Richard Simpkin, writing from the vantage point of 1980, almost a decade after *Marder* entered service:

> Since even the German officers concerned with its design and development cannot restrain their disappointment and criticisms of the Marder, they will be the first to appreciate what a disaster it looks to an outsider. With a weight of 25.5 t,[18] Marder is of an awkward density; it has marginal ground pressure—and thus track adhesion—when snorkeling (crossing submerged), and will not float without assistance. Its silhouette is massive and dangerously high for its class . . . around half a meter higher than Leopard 1 and 2. With reliability and ease of maintenance set as primary objectives, it is of almost tanklike complexity. . . . All in all the Marder project sticks up like a sore thumb in the Bundeswehr's otherwise, cost-effective and successful program of equipment for the land battle. It is more in tune with the story of equipping the Wehrmacht, or of the British Army of either era.[19]

The inherent drawbacks of the *Marder*, however pronounced, were less prejudicial than the Army's attitude regarding the characteristics of the new vehicle. As with the XM701 and its successor, the service was unwilling to trade off the characteristics of the armored personnel carrier as evolved over two decades—amphibious capability and sufficient capacity to accommodate a standard rifle squad—for the increased firepower and protection seen as necessary for an infantry fighting vehicle. Most of the perceived drawbacks of the German machine arose from the fact that the *Bundeswehr* had accepted these trade-offs. As a result, the alternative procurement strategy looked more promising; the M113, after all, was proving a great success. In some of its myriad roles in Vietnam, it was even carrying out missions arguably like those projected for the MICV, and the XM734 mounted-combat testbed (see chapter 3) had been quite well received. To this end, in 1967 the TACOM Program Manager, Combat Vehicles, awarded FMC Corporation, manufacturers of the M113, a contract for two prototype M113-derived vehicles, designated XM765.

The vehicle was duly delivered in spring 1969.

The XM765 was, essentially, an M113A1 (an M113 armored personnel carrier fitted with a diesel engine for greater fuel economy, fire resistance, and commonality with the future armored vehicle fleet) with approximately 1 ton of steel appliqué armor added to the front and sides, and a more powerful engine fitted to compensate for the weight. In addition, the features of previous vehicles were integrated. It used the XM734's layout of firing ports, but in a slope-sided troop compartment like the XM701's, the slope adding to the effective ballistic protection of the hull; the XM734's back-to-back seating arrangement for the squad members in the rear compartment was retained as well. A new weapons station, reminiscent of the cupola of the later-production M59 and capable of mounting either a .50 caliber machine gun or a 20mm cannon, was added as well, improving the weapon's usefulness in ground combat. It was, however, still operated by the commander, as in the M113. Finally, the rear ramp, retained from the M113A1, was flanked by a pair of armored fuel cells. These replaced the internal tanks of the standard vehicle, both enhancing safety by getting fuel out of the crew compartment and restoring volume—needed for storage and flotation—lost when the sides of the troop compartment were sloped. The fuel itself was diesel fuel; its higher flashpoint enhanced survivability still further.

The Army tested the XM765 at Fort Benning, Georgia; at Aberdeen Proving Ground, Maryland; and in Korea.[20] Although the tests were encouraging, the service did not proceed further with the vehicle, seeing it as insufficient improvement over the M113A1. More importantly, however, it was not an improvement over the XM723. Barring the fact that it existed and the XM723 did not yet, the XM765 was seen to sacrifice too much automotive and amphibious capability for the firepower and protection gained. Given the credence placed in the extensive conceptual studies, this result is not surprising. As one commentator has pointed out, "Whenever an actual piece of equipment is compared to an ideal standard, the hardware will be found lacking. . . . In conceptual studies, conceptual designs always work perfectly, while the shortcomings of actual vehicles are all too apparent."[21] The statement of one of the Army Combat Development Command Infantry Agency's equipment specialists is perhaps indicative of the Army's general position:

> Maybe many people are thinking that we could have the same capabilities with a product improved M113 Armored Personnel Carrier. After all, the M113 has been one of the most reliable tracked vehicle systems in the Army. Why go after a new vehicle and start learning all over again? It is true that the M113 can be product improved. The hull can be redesigned with sloped sides and squad firing ports. The BUSHMASTER gun can be mounted the same as on the MICV. Armor thickness can be added to provide the same protection as the MICV will provide. The power train can be improved to provide increased horsepower, and the suspension system can be improved to provide a better ride. But if we did this, what would we have? It would be like installing an eight-cylinder engine in a T-Model Ford; it's still a T-Model Ford.[22]

At the time of writing, of course, a thriving subculture of hot-rodders had been installing eight-cylinder engines in old Fords for a quarter of a century.

It is worth noting, however, that from 1970 on, FMC continued development of the XM765 as a private venture. This new vehicle, dubbed the Product Improved M113A1 (M113A1PI), while maintaining approximately 90 percent commonality of spare parts with the parent vehicle,[23] incorporated an uprated engine, a separate commander's station built into the cargo hatch, and an improved, laminated side armor developed by the Army's Ballistic Research Laboratory for the XM723. The M113A1PI, in turn, was further developed into a BMP-like configuration, with a one-man turret carrying a 25mm automatic cannon and a commander's station behind the driver. This became the basis for the Armored Infantry Fighting Vehicle (AIFV) used from 1978 on by the Dutch and Belgian armies, among others, and in a somewhat different form, closer to the original, by the Italians.

Having rejected shortcuts to an MICV, the Army proceeded apace with the original vehicle. The XM723 MICV was chartered as a program-managed item (i.e., an independent project) on 4 January 1968.[24] Tentative QMRs setting the vehicle's characteristics were submitted to the Department of the Army the following month.[25] In conformance with the Casey Board's recommendations, the requirements were revised to call for a smaller, lighter, and less expensive vehicle, and in this form the QMR was accepted in October.[26] Despite the numerous vehicle concepts brought forward in the Parametric Design Cost Effectiveness evaluation, which ideally sought to evaluate the weapons system in an unpreconcieved, ground-up fashion, the projected vehicle was still very much in the mold of the XM701 of five years earlier: an amphibious 12-man vehicle with a cannon mounted in a two-man turret,[27] to be protected against 23mm cannon fire from the front and 14.5mm from the sides and rear, a reduction from the Weapons Command's original goal of 23mm protection all around.[28] Projected to weigh 34,000 pounds, though, it was considerably lighter. Cost and weight continued to be factors, however, and as a result, a one-man weapons station was projected by the time of the program's July 1969 System Status Evaluation.[29]

By this time, then, the general characteristics of the XM723 MICV were fairly well defined. From the turret forward, the vehicle's layout was reminiscent of that of its Soviet counterpart, the BMP. The 450-hp diesel engine and the transmission were at the right front, with the driver beside them and the vehicle commander/squad leader to his rear. To the commander's right, behind the engine, was the gunner in his one-man weapons station, a turret mounting the Vehicle Rapid Fire Weapons System automatic cannon to one side and a coaxial M73 machine gun—the standard type used in postwar U.S. tanks, rather than the Infantry's M60—to the other. From the turret back, the MICV was unique among infantry vehicles to date, replacing the bench seats of other infantry vehicles with individual bucket seats for each of the ten occupants of the troop compartment in a crowded and somewhat bewildering arrangement. One man was seated facing forward immediately behind the commander and abreast of the turret, with two more in line behind him; two more were abreast of each of these two, but facing the rear of the vehicle; a final pair faced each other at the rear of the troop compartment on either side of the rear ramp. The seating arrangements were, in brief, arranged more with the idea of carrying the squad efficiently as passengers than with disembarking them efficiently as combatants. Along with an M113-style "cargo hatch" in the roof, these nine riflemen were provided with six firing ports, two on each side and two to the rear. They did not, however, use their own weapons from them. The riflemen were instead to use XM231 Firing Port Weapons (FPW) in much the same way as the *Marder* squad used the Uzi. The XM231 was a derivative of the standard M16 rifle, minus its front sight and buttstock. It was modified for sustained automatic fire by the substitution of a short, heavy barrel and by the adoption of open-bolt firing to prevent "cook-offs" (runaway firing caused by chambered rounds igniting in an overheated breech). The FPWs screwed directly into the ball-mount ports for better NBC security and steadier mounting. They were controlled, in the absence of sights, by observing the path of the tracer ammunition they fired. This arrangement was adopted as the result of comprehensive tests carried out by the Army Infantry Board. These showed that men standing in open hatches using their personal weapons observed about the same number of targets as those using vision blocks and firing port weapons, but that the specialized weapons could be brought to bear more quickly and with twice as many hits, aside from the benefits of maintaining overhead protection.[30]

Given the change to a one-man weapons station and the continuing concern over costs, yet another alternative to the XM723 as an all-new vehicle was investigated. The U.S. Army Tank-Automotive Command (TACOM) conducted a 1969–1970 design study for an "Austere MICV." This was to be a 17-ton vehicle (almost two tons less than projections for the XM723) using much of the running gear of the M551 Sheridan Armored Reconnaissance/Airborne Assault Vehicle (see chapter 6), and abandoning or deferring many of the sophisticated features of the XM723, such as night vision devices and gun stabilization. These economies realized significant cost savings. The "Austere MICV" was projected to cost $89,374, versus $151,575 for the XM723 as the design stood at the time.[31] Additionally, development time would be telescoped from 87 to 69 months. The austere vehicle's performance and capabilities, however, would have been at best a marginal improvement over the even cheaper M113A1-derived XM765 already in hand, which was being developed by FMC into the even more capable M113A1PI. The lost capabilities in night vision and ability to fire on the move would also open up the very differential in capability between tanks and mechanized infantry that the MICV program was supposed to address in the first place. The proposal thus went nowhere, especially since the M551 on which it was based was turning out to be problematic as well (see chapter 6).

Another 1970 evaluation approved a TACOM in-house vehicle concept, and by April 1972 Requests For Proposals (RFPs) were sent out to industry for test vehicles. FMC retained its preeminence in the infantry vehicle field, winning out in November over Pacific Car and Chrysler.[32] FMC delivered the prototype XM723 in December 1973, followed by sixteen engineering development vehicles over the next year. They emerged into a notably trying milieu. Events had transpired to raise questions as to the role of the tank, of that of the IFV in conjunction with the tank, and even of what sort of Army they would serve.

The October 1973 Arab-Israeli War had been a testing ground for a wide range of new technology and doctrine, as the belligerents deployed a panoply of the latest imported Soviet and Western materiel. On the Arab side, particularly with the Egyptian armed forces, the most prominent new equipment was a wide variety of guided missiles. These greatly influenced the conduct of the war, particularly in its early phases on the Suez canal front. The Israelis, building on the successes of the Six-Day War of 1967, relied on the combination of the fighter-bomber and the tank almost to the exclusion of other arms, consistently giving these forces priority in both modernization and mobilization.

Rather than face this threat head-to-head, the Egyptians used Soviet technology to augment their own armored and air forces. They covered their crossing of the Suez Canal with advanced surface-to-air missiles, which took a serious toll of Israeli attack aircraft before adequate countermeasures were fielded. Similarly, once across the canal, they were able to repulse the first Israeli armored counterattacks by using infantry armed with Soviet antitank weapons. The most important of these were wire-guided missiles with ranges of from two to three thousand meters: heavy, vehicle-mounted AT-1s and lighter and newer AT-3 Saggers. These were missiles of the type mounted on the BMP, but man-packed in suitcase-like carrier-launchers. These guided weapons were complemented by infantrymen armed with shorter ranged (less than 500m), unguided, RPG-7 (*Reaktivniy Protivotankovyi Granatomet*, rocket antitank grenade launcher) handheld antitank weapons. The Egyptian infantry were aided in their work by the nature of the initial Israeli responses, which were numerous, uncoordinated, company-sized efforts relying on tanks almost unsupported by infantry or artillery. The resultant loss of the ability to observe, clear, and direct suppressive fire on positions that might harbor antitank weapons made the tanks much easier targets for the massed fire of Egyptian missile and RPG teams. In this way, the Israeli quick-reaction force, the 190th Brigade, was almost wiped out.[33]

Much has been made of this episode, but it was not typical of the war as a whole. RPGs and wire-guided missiles had accounted for between 15 and 25 percent of Israeli tank losses, but the great majority, on both sides, were still lost to tank guns.[34] To be sure, the Israeli Armored Corps had sadly neglected its mechanized infantry elements, preferring to use the limited resources available to field more tanks. The infantry forces were largely mounted in a collection of M3 half-tracks left over from World War II and captured Soviet BTRs; these could not keep up with the tanks and were thus left behind in the initial effort to respond to the surprise achieved by the Egyptian army. Coincidentally, this motley fleet was just then being modernized with M113s, modified much as were the U.S. Armored Cavalry Assault Vehicles in Vietnam, with additional machine guns, and even 60mm mortars, for mounted combat.[35] As the Israeli Defense Force recovered from the initial surprise, completed its mobilization, and absorbed new materiel from the United States, it began to function more effectively as a combined arms team. As the war's lessons—and additional vehicles—were absorbed, the proportion of mechanized infantry squads to tanks increased to rough parity,[36] analogous to the earlier evolution of armored forces in other countries, notably the United States in World War II. Interestingly, the Israeli Defense Forces (IDF) did *not* develop or procure an infantry fighting vehicle. Instead, they relied on their fleet of M113s, called "Zeldas" in IDF service. These vehicles were enhanced with extra weapons, armor, and external fuel cells as on the XM765. In addition, the IDF converted redundant tanks to World War II–style "Kangaroo" configurations by removing their turrets.[37]

Arab mechanized forces, on the other hand, were nominally better equipped from the start; in particular, they had the BMP. The Egyptians claimed to be satisfied with the BMP, which they used to support missile-armed tank-killing infantry teams.[38] On the war's other front, however, the Syrians had less reason to be satisfied. Notably in the "Valley of Tears," in the Golan, the Syrians employed their BMPs in stereotyped armored attacks where their thin armor and cramped quarters left them dreadfully vulnerable to Israeli tank guns whose fire they were never meant to withstand.[39]

The Soviets themselves were alarmed by this turn of events. Their professional journals were filled with controversy, as various correspondents drew unsettling conclusions from the war. Chief among these was that their BMPs were the asset most endangered by the proliferation of antitank guided missiles (ATGM).[40] Tanks were heavily armored enough to resist some missiles' shaped charge warheads, and their vitals occupied a relatively small, out of-the-way area; the engine, fuel, and ammunition tend to be carried low in Soviet tank designs. Furthermore, their powerful high-velocity main guns gave them some chance to deliver counterfire against long-range missile attack. The BMPs, on the other hand, with their modest armor protection and large cargo of explosives and motor riflemen, would suffer grievous damage from almost any hit; their low-velocity guns had a much shorter effective range than even the crudest ATGM and were slower to respond due to the BMP's clumsy command arrangements. The Soviet development apparatus took these wartime lessons to heart when designing the next version of the BMP, the BMP-2, which emerged in 1981 with a two-man turret and a 30mm cannon.

The U.S. Army, then, found itself developing a vehicle to support tanks whose usefulness was suddenly more questionable than before. It was, furthermore, visibly derivative of one that had just acquitted itself poorly in its first wartime employment. In addition to the doctrinal uncertainty thrown up by the evidence of the Yom Kippur War, there was the problem of institutional uncertainty as well. The Army of the early 1970s was not a promising environment for a large new weapons system, especially a complex and novel one. The end of direct U.S. involvement in Vietnam in 1973 found the service in poor shape on practically every front. Morale sagged in the wake of the war. Relations with Congress and the media were poor. Demobilization portended reductions in force among the officer corps, and the end of the draft seemed to foreshadow a manpower pool of dubious quantity and quality at increased expense.

While adapting to this new personnel environment, it was meanwhile necessary to adapt to a new war-fighting emphasis as well. For the better part of a decade, the Army had been geared toward waging a largely counterinsurgency-oriented infantry war in Southeast Asia, albeit a complex, technology-intensive one. It now had to reacquaint itself with its relatively traditional areas of responsibility: the NATO theater and Korea. A force structure was in place to operate in these theaters, but it had been bled white of men and materiel, and showed the effects of neglect. Units were at modest fractions of their established strengths, and their equipment had not been substantially upgraded since the Kennedy administration had first reemphasized conventional land warfare. As a force heavily oriented to the demands of armored warfare in Europe, mechanized infantry suffered disproportionately from the new demands forced on the Army by Vietnam, especially in relation to the other infantry disciplines. As one lieutenant colonel, an Army War College student from an armored cavalry background, pointed out:

> We may well ask "What's happened to the mechanized infantry?" The requirement for straight-leg infantry in Vietnam and the increased use of airmobile operations there have contributed to mechanized infantry's present state.
>
> Infantry officers reaching the field grades today are oriented and trained for the airmobile assault or for dismounted combat. They may have had some exposure to mechanized infantry but, in many cases, it has been insufficient to provide the requisite base of trained, knowledgeable officers who can lead a mechanized unit now.
>
> While no statistics are cited here to support the contention that mechanized infantry suffers from a lack of qualified leaders, observation of the mechanized infantry battalions in training today confirms the contention. This lack of training and experience applies throughout the officer and NCO ranks. Furthermore, such training as is being accomplished is being done largely within mechanized units rather than in the schools.
>
> In this period of the Infantry's necessary preoccupation with Vietnam, the Armor officer has partially assumed the role of the mechanized infantry leader. By virtue of his training, which is devoted to the doctrine of combined arms and to the maintenance and management of heavy equipment, the Armor officer has been better prepared for the mechanized infantry role than his Infantry counterpart. The Armor officer has filled many of the vacant slots within mechanized infantry units. With respect to personnel, Armor today has a greater capability for employing and leading mechanized infantry than does Infantry.[41]

Similarly, modernization was postponed due to the demands of Vietnam, and such development of new systems as took place was slow; it must be noted that this problem was somewhat ameliorated by the presence of relatively new systems such as the M60 tank in the development pipeline. The delay was, again, partly due to the financial strictures necessitated by the problems inherent in fighting a war unbacked by a wartime economy. At the same time, however, the possibilities inherent in new technologies, especially electronics, had driven many new weapons systems to evolve into terrifically complex and expensive entities. Most of the electronic and electromechanical components that new ground weapons systems relied upon for their greater potential had been developed within the aerospace procurement and operating environment. The solid-state devices which made this advanced equipment possible were themselves quite reliable; however, their interconnections, the power supplies and cooling sysytems supporting them, and the actuators and transducers providing their inputs and outputs were often complex, finicky assemblies. As a result, such systems could be procured and maintained only with difficulty in modest numbers, given generous funding and careful maintenance by a plentiful supply of technicians under controlled conditions.

None of these parameters, however, could reasonably be said to apply to sensor, fire control, guidance, or communications systems operating in the Army field environment. This milieu involved thousands of copies of any given item, which had to operate in a largely adverse funding climate under the supervision of users with relatively modest maintenance skills in environmental conditions the Air Force or Navy would be likely to classify as untenably extreme and dirty. These conditions formed a set of rocks upon which Army weapons programs of almost every description foundered in the 1960s and early 1970s. Creeping complexity and its accompanying problems of cost and reliability claimed, among others, the Mauler antiaircraft missile, the AH-56A Cheyenne attack helicopter, and finally, the MBT-70 tank described above; more modest successor programs for all these had to be established to fill needs that continued to exist even though the programs set to fill them turned out to be untenable.

The XM723 MICV fared better than any of these programs, perhaps because, while automotively novel, its electronic advances were confined largely to its stabilized gun and modest night vision system. To be sure, its development was slow, but it did take place, and did so relatively publicly. As one of the few new armored warfare systems in prospect, it thus gained a good deal of attention from the user community, which began to speculate in the service press on how best to employ the new vehicle. Although its specifications were by no means finalized, the MICV was well understood in general to be an amphibious squad carrier of tanklike mobility, mounting an automatic cannon and fitted for the use of personal weapons while mounted, and protected to a higher level than that of the M113 APC, but by no means tanklike in its resistance to direct fire.

Some of these articles, especially those generated from within the Army development apparatus, regarded the MICV as more or less a "drop-in" one-for-one replacement for the M113 in the squad carrier role, and confined themselves to discussion of the new vehicle's capabilities in furthering that role.[42] In this they echoed official Army studies such as the 1971 IRUS (Infantry Rifle Unit Study), which saw the MICV-equipped mechanized infantry squad as identical to other infantry rifle squads save for the substitution of an MICV driver and gunner for two riflemen; it did, however, see scope for reduction in the overall size of all squads for better control by junior leaders and economy of manpower, due to improved personal and crew-served weapons.[43]

Officers outside the development community, on the other hand, took the potentialities latent in the MICV idea further, suggesting that its enhanced capabilities could be used to address some problems inherent in the current organization of mechanized and armored units. The ROAD concept of task-organizing combined-arms task forces from homogeneous tank and infantry battalions, as inherited from the U.S. armored divisions of World War II, had proven itself tactically sound, yet it was freighted with training and administrative problems. "Pure" battalions were simpler to run in garrison, and easier to train at the lowest level. At the same time, however, they made it difficult to form cohesive task forces and teams around battalion and company headquarters, the more so as it was necessary to coordinate training between multiple headquarters to form them, making for fewer opportunities to practice. In an Army constantly exhorted to "train as you fight," the contradiction was glaring. Task organization led to logistical complications as well: once task forces and teams were formed, the logic of cross-attachment meant that, for instance, a tank battalion headquarters might find itself with only one tank and two infantry companies, or an infantry headquarters the reverse, in defiance of the orientation and equipment of their supply and maintenance organizations. At the same time, junior officers in particular found themselves being evaluated by superiors other than those for whom they had done what was nominally their most important work. In one infantry major's words: "The problem previously has been that in combined arms task forces, the individual officers have normally been rated by their parent organizations, which may be miles away, and thus contact between the rated officer, the rater, and the indorser [*sic*] is rare."[44]

The prospect of the MICV, bringing as it did the capabilities and needs of small tank and mechanized infantry organizations more closely into line, seemed to open a way to address these problems by encouraging the creation of permanent combined-arms units. One of the most in-depth of these proposals available in the open literature was put forth in 1966 by then-Captain Dale K. Brudvig, an armor officer student at the U.S. Army Command and General Staff College. Captain Brudvig's thesis, incorporated into a College textbook on force design in 1971,[45] used official studies of World War II experience to show that armored combat commands, though theoretically ad hoc organizations, were in fact relatively consistent in their makeup.[46] He then used surveys of current practice under the ROAD concept to show that pure battalions were employed only about 8 percent of the time, and pure companies only about 24 percent.[47]

Finally, he used ROAD doctrinal manuals to show that "combinations of tank and mechanized infantry organize and conduct operations generally in a like manner whether a part of an armor or mechanized division."[48] To this end, Brudvig suggested replacement of the existing ROAD mechanized infantry and armored divisions with a single type of heavy division derived from the ROAD divisional base, but built around nine identical combined-arms battalions. Each of these consisted of two tank and two MICV-mounted companies of four platoons of four tanks or MICVs each, much as in contemporary French and Swedish practice. Brudvig's analysis showed that this divisional and battalion organization required slightly more manpower: 16,783 men versus 16,490 for the nominal ROAD armored division, the largest current establishment. However, it fielded as many tanks (324) as the armored division, and many more riflemen (3,456 vs. 2,709) than even the mechanized infantry division. Furthermore, it did so at somewhat lower cost through its elimination of extra headquarters and, perhaps most significantly, the elimination of mechanized infantry heavy antitank weapons such as ENTAC[49] wire-guided missiles and 106mm recoilless rifles, which Brudvig saw as superfluous in the presence of so many hand-held antitank weapons and tank main guns.

Several other officers carried this idea further in the pages of various professional publications, using persistent, if ahistorical, "dragoon" and "hussar" metaphors to advocate permanent combined-arms organizations down to company level—"dragoons" to be mechanized-infantry-heavy and "hussars" to be tank-heavy—as an aid both to combat training and to cohesion. None of these proposals, it is important to note, seriously proposed to eliminate cross-attachment entirely, only to reduce its frequency and adverse impacts.

In the meantime, the XM723s themselves were in as uncertain a state as the Army itself, even more undefined than most developmental vehicles. "Undefined" seems to be an odd term to apply to a 20-ton armored vehicle. The XM723, however, lacked much of its major combat equipment, however explicitly defined its general requirements had been. Its projected main armament, the "Bushmaster" cannon, or Vehicle Rapid-Fire Weapons System-Successor, had been in the offing for as long as the MICV itself, but was still lagging in development. Its caliber, long described in official documents as "20–30mm," had only recently been finalized at 25mm and reached competitive trials. As an interim measure, the 20mm M139 cannon fitted to the M114A2 reconnaissance vehicle was substituted. Similarly, the troop compartment's arrangement was by no means finalized, either in terms of the accomodation for the squad or even the weapons they fired. Once again, M3A1 submachine guns, the "grease gun" of World War II vintage, were fitted to the firing ports instead of the M16-derived weapons planned.

Nonetheless, the pilot vehicles sufficed for testing and the development of doctrine, and the program proceeded; novel arrangements were examined for their influence on human factors, and novel mechanisms, such as the innovative hydromechanical transmission, were examined for compatibility with service conditions.

In short, although the Vietnam War's demands reduced the resources available for MICV development, the development effort was emphasized following the discovery that the Soviet Union had already fielded an apparently analogous infantry combat vehicle. Extensive studies were made to determine the optimum characteristcs of the vehicle; these became a bone of contention as the various interested parties in the procurement process lacked a consensus on the desirable parameters for an MICV. The development process thus foundered as before, as the developers simultaneously insisted on enhanced combat capabilities and minimal logistical and fiscal impacts without compensatory design trade-offs being made; however, the development process continued. In the face of this impasse, efforts were made to determine whether foreign equipment or some development of the M113 APC would fulfill the requirement. These alternatives were rejected, once again, and for the same reason: developers insisted on greater capabilities at low prices without compensatory design trade-offs being made. Nevertheless, the development process continued. Events proceeded slowly, but relatively smoothly, until the MICV program collided with another one. As with the MICV, the effort to develop an Armored Reconnaissance Scout Vehicle had been troubled, similarly encumbered by an old complex of doctrinal and technological problems.

NOTES

1. "Does Theory Lead Technology?" *International Journal* 33 (summer 1978): 510.

2. *The Future British Surface Fleet* (Annapolis: Naval Institute Press, 1991), 7. The author was formerly chief constructor of the Royal Navy.

3. Isby, 123.

4. Richard Simpkin, *Red Armour: An Examination of the Soviet Mobile Force Concept* (London: Brassey's, 1984), 23.

5. The manpower pool provided by Soviet universal conscription made this restriction less onerous (Isby, 110).

6. This was not, however, entirely a measure of desperation; the diesel fuel provided a useful degree of neutron shielding when the tanks were full.

7. Master Plan, "Narrative," B. 2.

8. "Milestones." Usually rendered "Parametric Development/Cost Effectiveness Study" in other sources.

9. TACOM AHR'83, 223.

10. Bradley, "Future IFV," 26.

11. U.S. Army Weapons Command, *Technical Development Plan, MICV-70* (Rock Island, IL: U.S. Army Weapons Command, 1966), 5.

12. *Ibid.*, 7.

13. Kaufman, 186.

14. "Milestones."

15. Kaufman, 201ff.

16. *Ibid.*, 204ff.

17. *Ibid.*, 211f.

18. Metric, empty.

19. Simpkin, *Mechanized Infantry*, 30. The assistance Simpkin refers to is generally provided by a flotation screen, as on the Sherman DDs of Normandy vintage. This device is a continuous barrier erected around the vehicle, used to increase its displacement; it can be rigid, but is generally a collapsible fabric structure supported by a framework. This is the most common flotation aid for armored vehicles, though other approaches, such as detachable pontoons, have been tried.

20. R. M. Ogorkiewicz, "A New Version of a Well-Proven APC—The Product Improved M113A1," *International Defense Review* (April 1971), 164.

21. Kaufman, 190. For a parallel case in another branch see Edward C. Ezell, *The Great Rifle Controversy* (Harrisburg, PA: Stackpole, 1984), on the attempts by Army small-arms developers to displace the privately designed M16.

22. Jeff Cherry, "MICV: The Future Tool of the Trade for Mechanized Infantrymen," *Infantry* 62 (March–April 1972): 30. "BUSHMASTER" refers to the Vehicle Rapid-Fire Weapons System-Sucessor (VRFWS-S), a 20–30mm automatic gun intended to replace the 20mm cannon on light armored vehicles.

23. Richard M. Ogorkiewicz, "Infantry Combat Vehicles," *Infantry* 61 (January–February 1971): 49.

24. TACOM AHR'83, 219.

25. "Milestones."

26. TACOM AHR'83, 224.

27. Pre-IPR, "Justification for entering contract definition," "Chart #8."

28. *Technical Development Plan, MICV-70*, 6. This figure is interesting in that it suggests that WECOM expected Soviet thinking to move in the same evolutionary fashion as that of the United States, with the 23mm cannon replacing the 14.5mm heavy machine gun as a light vehicular weapon in the same manner as the still-undefined Vehicle Rapid-Fire Weapon System (VRFWS, or Bushmaster) was to replace the .50 caliber M2 and M85 machine guns. It also suggests a "mirror-imaging" planning mentality at work, envisioning mechanized infantry vehicles slugging it out with each other rather than fighting in combined arms. The requirement for 23mm protection was later completely dropped, as 14.5mm remained the norm for Soviet heavy machine guns.

29. *SSE,* "Technical Performance Narrative," 1.g.

30. Jeff F. Cherry, "Mounted Combat," *Infantry* 65 (September–October 1975): 13.

31. TACOM, *Austere MICV*, undated, 1. A copy of one edition of this document carrying corrections for the next, found at the Federal Records Center, Suitland, MD, gives some idea of the scale of the cost escalation problems associated with the MICV program; these costs were initally $83,637 and $124,995, respectively.

32. TACOM AHR'83, 224.

33. John A. English, *On Infantry* (New York: Praeger, 1981, 1984), 188.

34. Robin Rhoderick-Jones, "A Soldier's Tank," *Journal of the Royal United Services Institute*, 120 (September 1975): 32. Major Rhoderick-Jones at the time was serving in the Headquarters of the Director, (British) Royal Armored Corps.

35. Jac Weller, "Mechanized Infantry in Israel," *Infantry* 62 (September–October 1972): 34.

36. Jac Weller, "Armor and Infantry in Israel," *Military Review* 57 (April 1977): 10.

37. The IDF also stepped up development of a new tank, the *Merkava* (Chariot) designed for greater survivability and sustainability; one consequence of this design was a rear compartment normally intended for ammunition, but capable of carrying troops in much the same way as a pickup truck's bed may be used for passenger accommodation. Misinterpretation of this feature led to later confusion on the part of foreign observers, many of whom thought a sort of "super-IFV" had been invented.

38. Edgar O'Ballance, *No Victor, No Vanquished: The Yom Kippur War* (San Rafael, CA: Presidio, 1978), 161.

39. David C. Isby, *Weapons and Tactics of the Soviet Army* (London: Jane's, 1981), 127.

40. Phillip A. Karber, "The Soviet Antitank Debate," *Survival* 18 (May–June 1976): 106.

41. John P. Prillaman, "What's Happened to the Mechanized Infantry?" *Armor* 79 (September–October 1970): 29f.

42. See Jeff F. Cherry, "MICV: Future Tool of the Trade for Mechanized Infantrymen," *Infantry* 62 (March–April 1972): 28–30; and James F. McCluskey "Infantry Combat Vehicle," *National Defense* 58 (May–June 1974): 536–37 for examples. Cherry, a former infantry major, was an automotive equipment specialist with the Combat Development Command's Infantry Agency at Ft. Benning; McCluskey was Program Manager, MICV.

43. Robert J. O'Neal, "IRUS Recommends New Rifle Platoon/Squad for 1970–75," *Infantry* 62 (January–February 1972): 23.

44. James E. Mace, "The Combined Arms Battalion: Reality or Myth? " *Armor* 83 (July–August 1974): 22–26.

45. "A Proposed Divisional Organizational Concept," in *Selected Readings in the Design and Development of Combat Divisions* (Ft. Leavenworth, KS: United States Army Command and General Staff College, 1971), I:2-87–206.

46. *Ibid.*, 122.

47. *Ibid.*, 138f.

48. *Ibid.*, 142.

49. *ENgin Tactique Anti-Char*, a French-built lightweight antitank missile used by the U.S. Army as an interim weapon.

Chapter 6

Armored Cavalry: Problems of Doctrine and Equipment, 1940–1975

> One important difference between the US Army and the others we are considering is that they use designated main force combat teams for medium reconnaissance. This is avowedly because they consider the loss of speed and silence as less important than identification of reconnaissance force as such, and because they need fighting power to get information back with it. Maybe however the (in both senses) remarkably unhappy technical history of their tracked reconnaissance vehicles has influenced the American's thinking in much the same way as sustained lack of an adequate engine has affected British tank policy.
>
> —Richard Simpkin[1]

The complex relationship between armored forces and cavalry tradition was nowhere more apparent than in the history of armored reconnaissance. The U.S. Army's Cavalry branch had long been oriented toward reconnaissance and employment as mounted infantry, rather than shock action, given its experience in the Civil War and on the Great Plains. It is thus unsurprising that U.S. armored reconnaissance forces were not originally intended to engage in combat. Instead, they were expected to gather information by stealth. To this end, the divisional reconnaissance battalions of the Army's first armored formations were built around two reconnaissance companies. Each of these consisted of four platoons of four light scout cars each, plus a fifth platoon with sixteen motorcycles. These scouting forces were supported by a company of light tanks and one of armored infantry.[2] At the seminal 1941 maneuvers in Louisiana and North Carolina, the 1st Armored Division's 81st Reconnaissance Battalion, for instance, produced notable successes by stealth while incurring most of its simulated casualties through being lured into attacks on targets of opportunity.[3]

Wartime experience, though, was less clear. At Kasserine Pass, one company of the same 81st Reconnaissance Battalion so successful in Louisiana was surrounded and attacked, losing all but ten of its men.[4] At the same time, other missions arose for reconnaissance forces. In both North Africa and the European Theater of Operations, armored division commanders assigned their reconnaissance units a wide variety of tasks. These included attack, raid, pursuit, and advance and flank guard missions discouraged by official doctrine. A survey of divisional reconnaissance battalion commanders showed that reconnaissance accounted for only 13 percent of their employment by time; even much of that encompassed combat operations.[5] As with the later experience with the M113 in Vietnam (see chapter 3), designating an element of the force doctrinally as a noncombat asset did not necessarily serve to keep it out of fights.

The armored reconnaissance units were able to succeed at these demanding jobs in part by gaining strength, especially in their forward elements. As their parent armored divisions underwent successive reorganizations, the armored reconnaissance units continually gained in organic resources. By March 1942, even before U.S. forces entered battle the reconnaissance battalion had traded its rifle company for another reconnaissance company. At the same time, their component platoons were strengthened and homogenized: scout cars and motorcycles were replaced with armored cars and jeeps; these were supplemented with 75mm assault guns, at first mounted on half-tracks, then on light tank chassis.[6]

The following year brought one final reorganization, accompanying the definitive September 1943 reorganization of the armored division. In view of its enhanced combat power and range of missions, the reconnaissance battalion was renamed an armored cavalry squadron, and its companies were thus redesignated as cavalry troops, in line with traditional nomenclature. The nucleus of the squadron was the reconnaissance troops. These were now four in number, each armored cavalry troop consisting of three platoons. Each platoon was organized as one section of three armored cars and one scout section of three squads. Each scout squad in turn was mounted in a pair of jeeps, one armed with a pedestal-mounted machine gun and one with a 60mm mortar. The assault guns were removed from the platoon and brigaded into their own troop of eight weapons. The light tank company remained unchanged, making for an unusually large total of six maneuver and one headquarters troops under the squadron commander's control.[7]

Assessment of wartime experience led to a postwar revision of the divisional armored cavalry squadron in May 1948. To begin with, the squadron's name reverted to "reconnaissance battalion." The unwieldy span of control inherent in the existing squadron was remedied by changing the organization to a thoroughly homogeneous one. This consisted of a headquarters and service company and four identical reconnaissance companies, each of three identical platoons. Homogeneity ended sharply here; each platoon was a miniscule combined arms team in nine vehicles: a scout section mounted in four jeeps with machine guns and radios, plus another which carried the platoon leader; a section of two light tanks; a standard mechanized infantry rifle squad in its armored personnel carrier; and a support squad, manning a self-propelled 81mm mortar.[8] With appropriate changes of equipment, this platoon organization was essentially the standard until the late 1960s, spreading to the other divisional types with the adoption of the heavily armor-influenced ROAD concept.

Changes in equipment, though, were significant. Forecasts of the fluid, firepower-intensive, NBC-haunted battlefield of the late 1950s suggested that the unprotected jeep no longer had a place on the battlefield, especially as a mount for platoon leaders and company commanders. To this end, an armored command and reconnaissance vehicle project was set afoot in 1956, resulting in the M114 Armored Command and Reconnaissance Vehicle. The M114, a lightly armored, tracked amphibious vehicle, generally resembled a shrunken M113 with its boxy outline and top-mounted .50 caliber machine gun. It was, however, a separate design, manufactured by General Motors rather than FMC. Moreover, as one commentator has noted, a "more striking dissimilarity to the M113 was its lack of success in Army service."[9] Introduced in 1961, M114s were sent to Vietnam early, but were considered unsuitable for service due to deficiencies in their amphibious capabilities and cross-country mobility. They were replaced with M113s between April and November 1964, leading to another chapter in the larger vehicle's excellent record. Despite this experience, the M114 continued in Army service at home and in Europe, displaying indifferent mobility in the face of a considerable variety of terrains and upgrade programs. At last, in 1973, the vehicles were ordered withdrawn by the Chief of Staff, General Creighton Abrams. Only three thousand of the five thousand vehicles called for by the Army force structure were ever built.[10]

An analogous situation held with regard to the other major armored cavalry weapon, the light tank. By the end of the 1950s, the existing M41 light tank, with its high-velocity 76mm gun, was believed to lack the firepower necessary to counter the projected Soviet threat. At the same time, a requirement was developing for a light armored vehicle to replace the M56 "Scorpion" self-propelled antitank gun then serving in Airborne "tank" units. The M56 was the most austere vehicle imaginable in its role: it was little more than a small tracked chassis carrying a 90mm antitank gun, with only a small armored shield before the gun mount to protect the crew. It had been developed and adopted when the M41 had proven unsuitable for air-dropping. By 1961, the two sets of requirements were merged; they were to be fulfilled by the M551 "Sheridan" Armored Reconnaissance/Airborne Assault Vehicle (AR/AAV), a name that avoided the term "light tank" with its World War II connotations of ineffectiveness. The Sheridan was an effort to use advanced technology to reconcile the divergent requirements of airportability and decisive firepower in both the antitank and close support roles. Conventional guns compatible with a vehicle of the M551's combat weight—less than seventeen tons through the use of a hull fabricated from aluminum armor—lacked sufficient antitank capabilty. At the same time, traditional (to use the term loosely) antitank missiles were too specialized for other close support missions, such as firing high explosive and smoke rounds. The Sheridan therefore mounted a novel main armament, the XM81 152mm gun/launcher. This weapon, like that of the MBT-70, fired the Shillelagh guided missile as well as conventional ammunition; relatively conventional, that is, as these low-velocity rounds used a caseless propelling charge instead of the more usual metallic cartridge cases.

All this capability looked promising on paper, but the Sheridan's concatenation of novelties made for a troubled development history. The effects on the crew, as well as on the fire control and communications gear, of firing such a large weapon from such a small vehicle were only the most obvious problems. In addition, although the conventional ammunition had an effective range of around one thousand meters, the Shillelagh missile, due to the nature of its infrared command guidance system, had a *minimum* range of one thousand meters under the best of conditions. The lack of overlap was awkward and potentially dangerous, to say the least.

Furthermore, the Shillelagh had its own development problems. An automated beam-riding missile, the Shillelagh was, in its own right, as complex a front-line weapon as the Army had ever fielded. While undergoing tropical trials, for instance, it scored only two hits out of thirteen shots. At $13,000 per round, this was a problematic result, even disregarding the grave tactical consequences. It was all the more grave given that the Sheridan carried only ten of the bulky missiles.[11]

Nonetheless, the Sheridan went into production in 1966, and by early 1969 was fielded in Vietnam, replacing M113 ACAVs and M48 Patton medium tanks in various armored cavalry squadrons. Reactions to the new vehicles were mixed, partially depending on which vehicle it replaced. This was, in part, because the exchange of Sheridans for older vehicles was manpower based. Four-man Sheridans replaced M113 ACAVs, which had five- or six-man crews, on a three-for-two basis; the result was thus a significant gain in firepower and armor protection. When exchanged for M48s, on the other hand, the effect was to swap medium tanks for light ones, one-for-one, with obvious misgivings on the part of the users.[12] Nonetheless, some problems were noted by both sets of users. The Sheridan was prone to engine overheating and had a loud and unique supercharger whine which was especially undesirable in a reconnaissance vehicle; this signature alone could both disclose the Sheridan's location and identify it as well. Its fire control gear was found to be fragile and hard to maintain.[13] It was seen to be particularly vulnerable to mine damage in comparison to the M48. Most importantly, especially in light of the latter problem, the caseless 152mm ammunition was easily subject to damage under field conditions, and liable to be ignited if the vehicle was penetrated. On the other hand, under more favorable conditions, the firepower of the weapon was appreciated; the cannister round's effect was considered especially remarkable, though the missile capability was irrelevant to the Vietnamese combat environment. Additionally, the vehicle was considered superior in range, mobility, and night-fighting capability to both the ACAV and the M48.[14]

U.S. forces at home and in Europe appreciated the M551 more. In part, this was due to technical fixes arising from Vietnam experience, and in part because of the vehicle's greater suitability for Europe, where a temperate climate and a more benign operating environment better suited the Sheridan's capabilities. Most compelling of all was the growing perception of the M114's inadequacies with respect to mobility and reliability; eventually the Sheridan replaced the smaller vehicle in cavalry platoons in addition to supplanting the M56 self-propelled antitank gun and M41 light tank as originally planned. Less than five years later, however, it was in turn replaced in armored cavalry service by a mix of M60 main battle tanks and M113s, equipment that thus rendered armored cavalry forces almost indistinguishable from task forces formed from the maneuver battalions of armored and mechanized infantry divisions.[15]

The Sheridan was not, however, originally intended to serve as a replacement for the M114. As with any such program, studies for a successor vehicle were afoot even as the M114 was reaching the field. They continued slowly, marked by contention over roles and capabilities, until the M114's evident shortcomings forced hastier action. In October 1971, the Army's program manager for combat vehicles issued requests for proposals for the XM800 Armored Reconnaissance Scout Vehicle (ARSV).[16] Some idea of the magnitude of the bureaucratic task involved in getting the program underway can be seen in these remarks by Joseph B. Hayes, TACOM's chief engineer at the time. Because the problems here were common to other contemporary programs under discussion, such as the XM723 MICV, they are worth quoting at some length:

> Even before the Army's Source Selection Evaluation Board went to work in its judging of the Scout vehicle proposals offered by industry, the Commanding General of the Army Materiel Command decided it was time to review the source-selection procedures in the hopes of simplifying and streamlining them.
>
> In the Scout proposals, each offeror was limited to 1,250 pages. This was in response to a Request for Proposals from the Army in itself consisting of 935 pages. It appeared to AMC that any procedure which had to be limited to 1,250 pages in a proposal simply had too many unnecessary requirements in it. No doubt the same could be said about the RFP itself.
>
> The study group constituted to find a better system no doubt is knowledgeable from experience in source selection. Yet even experience may not provide a ready answer.
>
> Consider the problem of the developer of the Request for Proposals—usually the project manager for the weapon system. He starts with the relatively simple problem of translating a document called a Materiel Need (MN) or, earlier, a Qualitative Materiel Requirement (QMR) into a viable technical specification for the system.
>
> He then must address all the other requirements of a proposal. The outline of the management system of the offeror must be prescribed. His Integrated Logistic System

(ILS) approach must be delineated in specific areas. Finally, cost requirements input must be specified.

Each of these areas of the RFP takes months of soul-searching scrutiny on the part of the project manager and his staff, to ensure that he has fairly and adequately described the requirements without unduly limiting the offeror in exploiting new ideas and advancing technology.

When he then has in hand, hopefully, a salable RFP, he starts up the chain of authority for approval to put his program on the street.

This is where the ulcers set in. Although he is aware of the dollar constraints, the priority of requirements, and the problem of congressional mood, the project manager then must undergo a most frustrating series of briefings and justifications up and down the line of command and approval, before he finally reaches the one man who can say GO!

Then, more often than not, the GO man either doesn't say GO or bounces the program back for more detailed substantiation.

Finally, at long last, the project manager is given the green light to put his RFP on the street. Here he acquires his second ulcer. Along with all his experts, he had polished every detail of the RFP. It had to be perfect! It was—until six different offerors interpreted it six different ways.

Fortunately (perhaps) for the project manager, the system then sets him aside and places the evaluation of the proposals in the hands of a specially assembled group of experts known as the Source Selection Evaluation Board.

Now the going gets tough. A first analysis is made regarding the responses of the offerors in meeting the RFP. The board starts to examine each proposal in detail. No two are alike. And none is perfect; that is, trade-off studies show the need for compromises of requirement versus cost.[17]

Five manufacturers submitted proposals: the Allison division of General Motors (makers of the Sheridan), Teledyne-Continental (offering an Americanized version of the analogous Scorpion tracked reconnaissance vehicle, no relation to the M56, serving with the British Army), AAI Corporation, FMC, and Lockheed Missiles and Space Company.[18] Of these, the latter two were given contracts to submit vehicles for testing.

The two vehicles were radically different. The RFPs had specified a 17,000 pound amphibious vehicle, with enhanced mobility and sensor equipment relative to the M114, including a laser rangefinder, day/night fire control, and provision for a short-range ground surveillance radar. Once again, the "Bushmaster" Vehicle Rapid-Fire Weapons System-Successor (VRFWS-S) gun in a stabilized mounting was specified as the main armament. As with the MICV, the M139 20mm cannon, fitted as an upgrade to the M114A2, and no more successful than the rest of the vehicle,[19] was to stand in as an interim weapon until the VRFWS-S completed its own troubled development.[20]

Beyond this, the competitors had not been restricted in their approach, and indeed, their designs diverged sharply. Lockheed produced a wheeled vehicle with a novel six-wheeled, all-wheel-drive articulated chassis—that is, one in which the front half of the vehicle and its two wheels were pivoted to move freely relative to the rear half and its four wheels, an arrangement derived from an earlier Army testbed project, the 8-wheeled XM808 "Twister." FMC, on the other hand, turned in a more conventional tracked machine, visually reminiscent of the Stuart light tank of World War II vintage with its short overall length and high silhouette. Testing of the prototypes began apace, but by September 1974, the program was terminated. As with so many other Army procurement projects of the era—MBT-70, the AH-56 Cheyenne helicopter gunship, the Mauler air defense missile—increasing complexity brought costs without commensurate returns, which left the weapons systems vulnerable in the fiscal environment of the Vietnam and post-Vietnam eras. As General William DePuy put it, "Because of the cost of and preoccupation with the Vietnam war, the Army lost a generation of modernization."[21]

In addition, the Scout program was beset by fundamental uncertainties. Was the new vehicle to be a combat vehicle or sensor platform? If the latter, were all the lasers and radars necessary to the mission, or just the eyes and brains of the crew? Joseph Hayes, by then away from his post at TACOM, wrote an epitaph for the program:

> Why did Scout die? There are so many reasons that trying to pinpoint one or two would be a true exercise in futility. Yet let me try just two reasons.
>
> First, I believe, there was so much divergency of opinion on the need for Scout that there never was a solid requirement for the system. Rank held sway within the Army, and transfers and replacements of key officers led to a complete policy revision. The new leaders were finally in a position to control.
>
> Secondly, and certainly most important of all, is the lack of a workable system to define clearly to the development agencies and to industry the real requirements for weapons systems. It simply cannot be left to the vagaries of opinion of one general replacing another at or near the top of the Army Christmas tree. Somehow there must be evolved a system for conducting policy on requirements.[22]

The impracticability of procuring the ARSV did not, however, eliminate the need for a scout vehicle. To this end, a new search was mounted for a nondevelopmental substitute, covering vehicles ranging from motorcycles and dune buggies to the Sheridan and the XM723 MICV. The latter was itself in danger of succumbing to the same forces of role confusion and cost escalation that killed the MBT-70 and the ARSV, and was using so much of available development resources that a new scout vehicle was out of the question. It appeared, then, that only one vehicle could be procured in the existing fiscal and political climate. As a result, by the end of 1975 efforts got underway to meld the MICV and Scout requirements, with their disparate characteristics, into a single vehicle. Given the history of the separate predecessors, the task was daunting.

In summary, then, simultaneously with the drawn-out problems associated with the MICV program, a somewhat analogous crisis was occurring in the armored cavalry community. Armored reconnaissance had been conducted since 1940 in a slippery doctrinal environment with a variety of makeshift and experimental equipment, none of which was very satisfactory. In the 1960s, an effort was gotten underway to develop and procure a specialized Armored Reconnaissance Scout Vehicle (ARSV). This development program collapsed in the mid-1970s, as developers insisted on greater capabilities at low prices without compensatory design trade-offs being made, in a manner highly analogous with the previous problems of the MICV. In an effort to procure a new scout vehicle, the development process of the fiscally troubled and doctrinally unstable ARSV program was merged with that of the fiscally troubled and doctrinally unstable MICV program.

NOTES

1. *Tank Warfare* (London: Brassey's, 1979), p.65.

2. I. D. White, "The Armored Force, Reconnaissance Battalion, Armored Division," *The Cavalry Journal* (May–June 1941), quoted in Peter S. Kindsvatter, *The Army of Excellence Divisional Cavalry Squadron—A Doctrinal Step Backward?* (Ft. Leavenworth, KS: School of Advanced Military Studies, U.S. Army Command and General Staff College, 1985), 42.

3. Peter S. Kindsvatter, *The Army of Excellence Divisional Cavalry Squadron—A Doctrinal Step Backward?* (Ft. Leavenworth, KS: School of Advanced Military Studies, U.S. Army Command and General Staff College, 1985), 6; hereafter given as *AOE Cav*. Then-Major Kindsvatter was a student at SAMS.

4. Kindsvatter, *AOE Cav*, 6.

5. General Board, United States Forces, European Theater; "Tactics, Employment, Technique, Organization and Equipment of Mechanized Cavalry Units," Study No. 49, 4 February 1947, Appendix 4, p. 2; quoted in Kindsvatter, *AOE Cav,* 8f.

6. Kindsvatter, *AOE Cav,* 42.

7. *Ibid.,* 43.

8. *Ibid.*, 44.

9. Zaloga and Loop, *Modern American Armour*, 31.

10. Brooke Nihart, "Army Moves Out on New Armored Vehicles, Gun," *Armed Forces Journal International*, 108 (3 May 1971): 34.

11. Zaloga and Loop, *Modern American Armour*, 32.

12. Starry, *Mounted Combat,* 142f.

13. Zaloga and Loop, *Modern American Armour*, 32.

14. Starry, *Mounted Combat,* 144.

15. Wagner, "Broken Saber," 36.

16. Hayes, "Tracks or Wheels," *Ordnance* 51 (January–February 1972), 274.

17. Hayes, "Selection for Scout," *Ordnance* 51 (May–June 1972), 450.

18. Nihart, "Army Moves Out," 34.

19. Zaloga and Loop, *Modern American Armour*, 32.

20. Joseph B. Hayes, "The U.S. Army Scout Vehicle Competition," *International Defense Review* 6 (February 1973): 86.

21. DePuy, "U.S. Army," 24.

22. Joseph B. Hayes, "Scout Bites the Dust," *National Defense* (July–August 1974): 26.

Chapter 7

From MICV to Bradley Fighting Vehicle, 1975–1983

> There is not much combat effectiveness in a requirement document that calls for a better weapon system than the enemy already has fielded.
>
> —Brigadier General (Retired) Philip L. Bolté
> Former TACOM Program Manager,
> Fighting Vehicle Systems[1]

The collapse of the Armored Reconnaissance Scout Vehicle project and its subsequent merger into the Mechanized Infantry Combat Vehicle program late in 1975 greatly increased the importance of the latter to the Army. It also introduced an entire new level of complexity into the XM723 MICV's development.

Although the Scout and Infantry variants of the MICV were still automotively identical, they were envisaged as having different weapons stations. The infantry version was to continue with the one-man turret as planned; not one, in fact, but two of them. The planned cannon armament, the Bushmaster, or VRFWS-S was still in development—indeed, the program had not yet been narrowed down to a single candidate weapon. Therefore, an interim weapons station mounting the M139 20mm cannon continued in development. Despite the lighter weapon's disappointing performance to date on the MICV-65, the M114A2, the M113A1PI, and the two ARSV competitors, the risks incurred by the delays in the Bushmaster program were so great as to outweigh the M139's historical baggage.

The Scout vehicle was a different matter. Its reconnaissance mission placed a premium on observation for the commander. The original MICV arrangement, with the commander stationed in the hull behind the driver and beside the turret, was unacceptable. The Scout was thus to have a two-man turret, so the commander could be stationed at the highest point in the vehicle, with a 360-degree field of view. Interestingly, when the Soviets adapted the BMP for their armored reconnaissance units to replace the PT-76 light amphibious tank, they made the same change. This revision meant a larger turret ring (60 inches versus 42 inches), so it imposed structural changes on the hull as well.

The armored reconnaissance mission also demanded changes in internal layout. The evolving Scout mission called for a wide variety of surveillance and communications equipment, including such items as a motorcycle and a small ground surveillance radar. At the same time, projected combat roles suggested that the vehicle needed additional ammunition.

Ammunition stowage was a particular issue because, in addition to the cannon and coaxial machine gun of the MICV, the Scout variant was to mount the TOW (Tube-launched, Optically Tracked, Wire-guided) heavy antitank missile, a sophisticated 3,000-meter range, automatically wire-guided weapon.[2] Previously used by infantry units and armed helicopters, addition of the TOW to the scout vehicle represented a departure from previous armored cavalry practice. Heretofore, heavy weapons had been kept separate from scout vehicles to maximize stealth and emphasize the information-gathering mission. This departure derived from the Army's post-Vietnam reorientation to European warfare, which foresaw a need for antiarmor firepower in forward areas to counter Soviet numerical superiority. This concept was soon to grow into the "Active Defense" doctrine of 1977.

Quite aside from the administrative and engineering complexity involved, this situation was insupportable. In a time of fiscal retrenchment and reductions in force, asking Congress for funding to produce a single vehicle with two hulls and three turrets seemed the quickest way to have the project join the MBT-70 and the ARSV. The embarrassment latent in this situation was all the greater given that the Army had dismissed the possibility of an M113-based MICV years before, due partially to the undesirability of a mixed fleet. To address these problems, in August 1976 the Department of the Army appointed Brigadier General Richard Larkin, assistant commander of the 4th Infantry Division (Mechanized), to head a task force to examine the MICV program. After three months' deliberations, the Larkin task force recommended that development of both one-man weapons stations be abandoned. Instead, it was recommended that an essentially common vehicle with the same level of protection as the existing vehicle should be procured. This course of action was reinforced by the perception that the commander's limited visibility and separation from vehicle's weapons and sensors had been a major shortcoming of the BMP in 1973.

Combat experience was confirmed by experimental evidence. While the Larkin task force was sitting, human factors tests were carried out in the course of second-phase operational testing. These tests found that the XM723 MICV, which shared the BMP's layout, was equally difficult to command, despite the addition of a light-emitting diode display to help the commander transmit target bearings to the gunner in the turret.[3]

Under the Larkin task force's scheme, the cavalry vehicle would thus mainly differ from that of the infantry in carrying a smaller crew and more ammunition, while omitting the firing port weapons. In both roles, the vehicle was to mount the turret previously associated with the scout version. This was still in an early stage of development, but by now was designated TBAT-II, for TOW-Bushmaster Armored Turret, Two-man.

In some ways, this innovation, which had been discussed within the service for some time,[4] was revolutionary. It put the TOW, which had previously been regarded as a company- and battalion-level antitank weapon operated by specialists, at the disposal of each mechanized infantry squad. It thus gave mechanized infantry units a density of antitank firepower unheard of since 1942, when the *Wehrmacht*'s 90th Light Division, part of the *Afrika Korps*, had issued a captured Soviet 76.2mm field gun to each squad.[5]

In other ways, the addition of TOW was an evolutionary step. Mechanized infantry forces had, after all, traditionally deployed large numbers of antitank weapons, both to provide a positional anchor to support armored maneuver and to free up tanks that would otherwise be required for their protection in the defense. In that tradition, for instance, the new and somewhat problematic Dragon antitank missile was already assigned at the mechanized infantry squad level. This lighter weapon was a 1,000-meter-range missile, also with automatic wire guidance, nominally carried and fired by one man.[6] It had rudimentary provisions for firing from the M113, albeit not from under armor. The proliferation of these antitank missiles was thus analogous to deployment of the bazooka, which had been issued to each World War II armored infantry squad, rather than at the company level where it originally resided in other infantry tables of organization and equipment. Indeed, the proliferation of antitank guided missiles is strikingly parallel with that of the machine gun, which went from being a battalion weapon to a squad weapon over the course of World War I, and from a squad to a fire team and even (in the somewhat different guise of the assault rifle) a personal weapon during World War II.

Significantly, both TOW and Dragon were definitely infantry weapons in at least one respect: they had to be fired from a stationary position, which had to be maintained while the subsonic missiles flew to the target. This tended to dictate long standoff ranges, allowing the long-range accuracy of the missiles to compensate for the faster reaction time and higher rate of fire of opposing tanks. Applied to the use of TOW and Dragon by the redesigned MICV, however, this meant that vehicles in the most advantageous positions to use their missiles would likely not be well positioned to support dismounted mechanized infantrymen clearing obstacles or defiles in support of tanks. Although defensible, this tactical dichotomy would come back to embarrass the Army later, the more so because of the disparity in ranges of the various weapons aboard the vehicles. The TOW ranged out to 3,750 meters, with a premium on long range, while the 25mm was good to less than 2,000 meters, even discounting the much shorter ranges of the weapons carried by the dismount teams. The situation was further complicated by the prospect of deploying the new systems in Western Europe, where urbanized and forested terrain made anticipated engagement ranges much shorter, often 600 to 1,000 meters.

Whether revolutionary or evolutionary, the addition of TOW to the Cavalry Fighting Vehicle at least had a firm basis in doctrine. In the summer of 1976, just before the Larkin task force began its deliberations, the Army Training and Doctrine Command (TRADOC) under General William DePuy had issued a new edition of the Army's key doctrinal manual, *FM 100-5, Operations*. This work marked a radical reorientation of the Army's professional thinking from the problems of peripheral war and counterinsurgency in general, and Vietnam and its aftermath in particular, to those of central war in Europe against the forces of the Warsaw Pact.

As discussed above, NATO warfare presented serious problems, quite aside from those of the defenders' numerical inferiority. The politics of Cold War Europe precluded some of the measures to which a numerically inferior defender might otherwise resort. Fortification of the Inter-German Border would have signaled tacit acceptance of the country's partition. The exchange of West German territory for time to build up reinforcements would inflict terrible damage on the country, even if there were sufficient territory to yield the necessary time. Preemptive attack was surely out of the question in the post–World War II European diplomatic system. At the same time, the possibility of conventional operations escalating into tactical or even strategic nuclear warfare meant that the quick, decisive repulse of a Warsaw Pact invasion was imperative.

Drawing heavily on the experience of the Yom Kippur War, DePuy saw the firepower of the new generation of weapons as providing a palliative, if not a remedy, for this situation. In particular, at the tactical level he envisioned a defensive screen of "covering forces," well-supplied with antitank weapons—armored cavalry units equipped with TBAT-II-configured scout vehicles, for example. These units were to force the attacking formations to stop, deploy, and engage them, thus incurring delays and telegraphing plans. The time and intelligence thus gained would in turn allow the main battle forces—tank- and mechanized infantry-based task forces—to redeploy and concentrate, while air and artillery firepower from the rear battle area would be brought to bear as well. This marked a shift of emphasis in the armored cavalry's doctrinal role, if not its actual one (see chapter 6), tending to integrate the combat and information-gathering roles that had previously been viewed as divergent, if not mutually exclusive.

DePuy also saw the MICV as a key element in the conduct of this "Active Defense." It was necessary "to support tank-led combat teams by: long-range suppression of enemy anti-tank weapons, or suppression of the same enemy capability while the MICV is moving cross-country with tanks . . . or delivery of a high volume of close-in overwatching suppressive fire in support of dismounting infantry . . . defeat the BMP . . . and be able to fire an ATGM from the deck,[7] and protect against automatic weapons fire." DePuy went so far as to deem the MICV an "issue that goes to the heart of the Army's capability."[8]

The TOW, then, was seen as an emerging doctrinal necessity for the scout variant. For its infantry counterpart, the addition of the missile was more important politically than doctrinally; a common hull and turret for the two vehicles made for a vital savings in development costs. Furthermore, without it, the MICV, whatever its virtues, represented only an incremental improvement over the M113 and its derivatives. The total cost increment for each vehicle equipped with TBAT-II as opposed to the basic XM723 design was given as $79,000 (Fiscal Year [FY] 1976) in Congressional testimony. It is worth noting that much of this cost was for the thermal sight, an item deemed necessary in any event for both the scout mission and to match the night-fighting capability of the XM1 main battle tank.[9] For this relatively small monetary outlay, the XM2 gained the new guided missile capability that had made such an impact in the early phases of the Yom Kippur War, and did so without requiring the development of a new weapons system. Moreover, it did so by putting the weapons under armor, rendering them resistant to the suppressive fire used as a counter to antitank missile attacks in 1973. Previous mobile ground mountings for the TOW, on M151 jeeps and atop M113s, had been unprotected. As General Donn Starry, DePuy's successor at TRADOC, stated: "We in TRADOC . . . decided to put the TOW on the MICV because we realized that if we did not put the TOW on the MICV, we would probably never have an MICV."[10] Even without this rather pragmatic rationale, the Army's addition of TOW to the MICV made considerable sense in terms of the service's available resources. Time and fiscal constraints left the Army with the twelve-man XM723 hull and automotive systems on its hands, with its development almost complete. These were sized for a mechanized infantry squad the Army would likely be unable to field, within the constraints of the post-Vietnam volunteer Army. With only the gunner in the turret and the commander/squad leader in the hull, the XM723 was, in any case, configured in a manner both combat experience and experimentation suggested were at best inefficient and at worst untenable. Assuming the necessity for staying with the MICV hull and power train already developed, the TBAT-II design was a useful exercise, in effect taking up the excess volume in the MICV's design while correcting the problematic configuration and addressing the perceived Soviet armor threat. A missile avoiding the mobility penalties inherent in the TOW would have been better, and would certainly have made a more clear-cut argument for antitank capability for the new vehicles, but no such weapons system was available.[11]

Preliminary work on the modifications necessary for the TBAT-II configuration having already begun, the redesign, based on the MICV Task Force's detailed recommendations, went ahead quickly. Larkin's group, cognizant of the increased level of complexity and expense this entailed, recommended that the TOW installation be relatively simple, carrying two rounds rather than the four originally considered, in a nonelevating mount, unlike those being studied for specialized M113-based TOW carriers. Similarly, they recognized the probability that the new turret would drive up the vehicle's weight. In view of this problem, the Larkin task force recommended that the MICV's amphibious capability be achieved by use of a water-barrier system. This was a bathtub-like canvas enclosure around the hull to increase waterborne displacement, as pioneered by the Sherman DD's of World War II and currently used by the Sheridan. Use of a water barrier was probably inevitable if amphibious capability was to be achieved, as the MICV's hullborne flotation had become marginal at best, even with the lighter one-man weapons station.[12]

Perhaps the most significant design impact of the new arrangement was on the squad. The original MICV was sized for the then-current eleven-man infantry rifle squad standard throughout all types of infantry unit, whether mechanized, regular, airborne, or airmobile. Only six of these men, however, aside from the vehicle crew and squad leader, were placed where they could observe or fight from under armor. Accommodating the larger turret ring and providing stowage for reserve missile rounds effectively eliminated the odd two men. Officially, a trend was emerging to reduce the size of the squad to nine men in view of manpower concerns; furthermore, Army research such as the IRUS of the mid-1960s had suggested that improved firepower allowed a squad as small as seven. At any rate, in practical terms, few Army infantry squads at this time were ever at full strength. Even under the relatively improved conditions of the early 1980s, many M113-equipped mechanized infantry platoons could field only eighteen to twenty-five men, making for de facto rifle "squads" of five to seven which concentrated on manning their most potent weapons first.[13]

FMC was quickly awarded a contract to produce the modified vehicles, and the redesign effort continued. In March 1977, the MICV program was renamed "Fighting Vehicle Systems" in recognition of its wider purview, as well as of the effective termination of the 20mm-armed MICV on which development had previously been concentrated. Similarly, in May the departure from the original XM723 MICV concept was recognized by yet another redesignation: the MICV became the XM2 IFV, while the Cavalry vehicle became the XM3 Cavalry Fighting Vehicle (CFV). The following month, the program's scope was widened even further, as the propulsion and suspension systems of the M2/3 were designated as the automotive basis of the Fighting Vehicle Systems Carrier (FVSC). This was to be a general-purpose armored load carrier, initially intended to mount the General Support Rocket System (later designated the Multiple Launch Rocket System),[14] a heavy unguided-rocket launcher originally intended to provide counterbattery fire for division and corps artillery in place of the M110 8-inch self-propelled howitzer. The redesign progressed apace, as several of the thorniest problems of the original MICV effort were worked out. In particular, the advanced hydromechanical transmission finally appeared to be satisfactorily living up to its potential for improvements in fuel consumption and maneuverability.[15] General Electric, the original contractor, won the contract for this component even after a competing Allison design was introduced into the competition.

All this redesign, while relatively straightforward, took time and raised costs. This, in turn, drew congressional attention. Some legislators, notably Senators Sam Nunn (R-GA) and Gary Hart (D-CO), as well as Hart's defense aide, William S. Lind, questioned the entire rationale for the MICV, given its roots in the strategic and technological environment of the early 1960s. They believed the system was already obsolescent in concept, in particular due to its limited protection relative to the XM1. They favored, instead, a heavier infantry fighting vehicle with tanklike levels of protection. The Army had preliminary studies for such a vehicle under way, but appears not to have favored such a solution on the grounds of cost and weight, which would have been on the order of those of the M1 tank.[16]

As a result, hearings were held in March 1977, before the Research and Development Subcommittee of the Senate Armed Services Committee to examine the program. Because this hearing encompassed many of the issues that would later bedevil the M2/3 program, it is worth examining at some length. The Army's main witnesses were General William E. DePuy, commander of TRADOC, and the Hon. Edward A. Miller, assistant secretary of the Army for research and development. General DePuy opened the testimony with a discussion of the strategic situation in Europe, pointing out the political need for forces in West Germany to defend as far forward as possible. Given the limited forces available, Depuy asserted, this situation placed a premium on highly mobile forces capable of rapid concentration to meet Warsaw Pact armored thrusts, fielding enough antiarmor firepower to stop such thrusts. In turn, it was necessary to provide these antiarmor weapons with sufficient physical protection to allow them to function in the face of the intense artillery concentrations expected to support Soviet-style attacks, which even in World War II could reach densities of up to 350 tubes per kilometer of front.[17] The foregoing, although not named as such, was the rationale for the "Active Defense" doctrine promulgated by DePuy the previous year. It is worth remarking that William Lind was a noted critic of the doctrine, which he described as "garbage"[18]—or, as one TRADOC historian put it, "seriously deficient"[19]—for its excessive reliance on attrition.

DePuy then went on to outline a generic battalion engagement between U.S. and Soviet forces as currently equipped, showing how tank, mechanized infantry, and TOW platoons interact, and emphasizing the limitations placed on the M113-based infantry and antitank forces due to their shortfalls in mobility and protection relative to the tank:[20]

> Right now our infantry has a very poor vehicle. Why? First of all, if you get in it and ride in it, everybody agrees the M-113 is not very good. It's very difficult to move around. Second, the 50 caliber machine gun is not a good suppressive weapon. Third, the commander wants infantry to accompany the tanks in attack, because when the tanks get into a little village on the hill they need infantry. Thus, you need to be able to send the

infantry with the tanks.

If you sent the infantry with the tanks, you would like to have them do the suppressive job while moving, and that is why we have put a cannon on the MICV. That is why we have stabilized it. And we then have gone one step further. We can talk about this later. As long as we are going to sink the money in a vehicle that can keep up with the XM-1, that can fire suppressive cannon while on the move, and can turn back 50 caliber machinegun fire at fairly close range, and since we really want to put more antitank guided missiles on the battlefield, let's put some antitank guided missiles on the MICV because the incremental cost is practically nothing. The incremental cost is very little and most of it is simply, the value of the tracker and the value of the missile itself.[21]

General DePuy's argument for the MICV was followed by an extensive, although somewhat confused, discussion of the design of the MICV. Committee members raised a variety of questions about the vehicle's layout, in particular its height relative to both the XM1 and the Soviet BMP infantry fighting vehicle. Miller explained the various items contributing to the 112-inch total height of the vehicle, the largest contributors being the need to provide sufficient wheel travel for enhanced cross-country mobility, the need to accommodate American soldiers of ninety-fifth percentile height, that is, well over six feet tall—a problem not shared by the designers of the BMP (see chapter 5)—and the need to provide ten degrees of depression for the gun.[22] DePuy used this last datum to enlarge on the differences between Soviet and U.S. doctrine as expressed in design philosophy. U.S. doctrine called for vehicles to make the maximum use of terrain features for protection and concealment; as a result, overall height was sacrificed for main gun depression. The Soviets, on the other hand, tended to minimize tactical maneuvering at the lowest levels, which called for a lower vehicle to reduce target area when in the open.

The inquiry then turned to the XM2's strategic mobility, in particular, the need to remove its side armor arrays to allow the vehicle to fit into the C-141 Starlifter for air transport, one of the very parameters on which the MICV-65 program had foundered over a decade earlier. The testimony on this issue revealed the confusion inherent in the prevailing ideas about the MICV concept. The Senators' questioning bogged down in extensive and minutely detailed questions about the time constraints and loss of protection without asking about the value of the vehicles in the absence of the tanks they were intended to work with, which were not portable in the C-141, and at the time barely portable in the scarce and technically troubled C-5A. Equally remarkable was the failure of the military witnesses to point out that, however cumbersome the process of removing and replacing the side armor for airlift was, it was still incomparably faster than sealift.

Perhaps the most striking characteristic of the Army's testimony in these hearings was the essentially negative character of the defense of the MICV. General DePuy, who nominally saw the the MICV as fundamental to his new doctrine, still tended to couch his case for the system in terms of sunk costs, justification by foreign practice, and the inadequacy of the M113 for mobile armored combat, rather than any inherent quality of the vehicle itself. For instance:

> . . . almost every army you look at is ahead of the American Army, as far as taking care of our infantry. The Russians are ahead of us, the Germans are ahead of us, the Dutch are ahead of us, the French are ahead of us, the Yugoslavians are ahead of us. Almost everybody has a better infantry vehicle than the U.S. Army
>
> We would have been better off in 1963 when we started to just build the MICV immediately, and we would have had it. It would not have been perfect. But an alternative that we have, it seems is to reverse engineer[23] a BMP. I am saying we do not want it. The Marder is not as good. It would cost us more and weighs more . . . . Yet, it's in the hands of a good army and they have got it. Are we to start over again? My guess is that if you start over again, you will get a 10-percent increase in effectiveness and a 50-percent increase in cost.[24]

Brigadier Richard Simpkin's lampoon of the Army's attitude—"Well, I guess we have to have this MICV, 'cos everybody else has one so otherwise we'll be disadvantaged"—is an exaggeration, but not by much.[25]

In the wake of these hearings, Congress directed the Army to examine the IFV concept further, with an eye to the costs involved. In October 1977, therefore, yet another Department of the Army task force, this time chaired by Major General Pat Crizer, reevaluated the requirement for the IFV and CFV, examining their design and making inquiries into the need for a more survivable vehicle.[26] Congress did not, however, use the Department of the Army as its sole investigatory tool. It also directed the General Accounting Office (GAO) to investigate the Fighting Vehicle Systems program, especially with respect to the vehicles' compatibility with the XM1 tank, itself in development at a somewhat more advanced stage, and itself the target of intense congressional and press scrutiny.

A brief discussion of the XM1 is thus necessary here.[27] Escalating costs and complexity had led to the cancellation of the MBT-70 in 1970. The following year Congress cancelled the XM803, an "austere" version of the MBT-70 put forward in an effort to salvage the program. Because the failure of the program did not vacate the requirement for a new tank to replace the M60, Congress authorized a new main battle tank program, to be conducted under strict unit-cost limits.

Although the Army developers maintained that they had taken a cautious approach, the vehicle that had emerged from the new program, the Chrysler Corporation's design for the XM1 (later named the Abrams), was as great a departure from previous practice as the MBT-70 had been; it was, however, more subtly so. It may be argued that in most of the ways that the MBT-70 was revolutionary, the XM1 was evolutionary, and vice versa. The MBT-70's layout was radical; the XM1's was traditional, with the driver in the hull and the main gun served by a human loader. The MBT-70's enhanced firepower was obtained through a novel armament; the XM1 carried much the same armament as its predecessor, but got more from it through advanced digital fire control, a gunner's primary sight employing an imaging infrared sensor, and much more effective main gun stabilization. The MBT-70 was powered by a traditional diesel engine (albeit one with baroque performance enhancements such as a variable compression ratio) but it rode on a novel hydropneumatic suspension; the XM1 was the first tank to be powered by a helicopter-style gas turbine engine, but rode on torsion bars, albeit ones forged from a new high-strength steel. Finally—and perhaps most significantly—the MBT-70 sported advanced protection against NBC threats, but offered little improvement in armor protection; the XM1's NBC defenses were essentially those of the M60, but its ballistic protection was based on a new composite armor system derived from the advanced, highly classified British "Chobham" armor. This two-foot-thick laminate rendered the tank apparently invulnerable to ground-launched antitank missiles over the frontal arc, and against almost all gun attack as well; even if penetrated, the XM1 incorporated an advanced fire suppression system and innovative ammunition storage arranged so that gases from secondary fires and explosions—historically, the source of most catastrophic tank kills—would be vented overboard with relative impunity, through a set of blow-out panels in the hull and turret roofs.

It is worth noting that many of the sophisticated features of the XM1, however valuable on the armored battlefield, tended to increase the need for a capable IFV. The need for future infantry vehicles to match the new tank's cross-country performance was obvious. More subtle factors came into play as well. The very turbine engine that contributed to the XM1's mobility made it difficult for infantrymen to use the tank's armored bulk for cover, due to the intense heat of the exhaust. The XM1's very excellence as a tank-killer also got in the way of combined-arms operation. The idea that "the best antitank weapon is another tank" had reached the status of a truism in the U.S. Army since World War II and the failure of the tank destroyer doctrine. Nonetheless, the postwar line of medium tanks—the M46, M47, M48, and M60—had maintained considerable capability to provide close support to infantry against nontank opposition. The XM1, however, was more specialized for antiarmor warfare, and was projected to become more so in the future when fitted with a 120mm gun as its main armament in place of the original 105mm weapon. This would find the new tank with a smaller supply and restricted variety of main gun ammunition. The 25mm gun of the M2, with its copious supply of high-explosive and armor-piercing ammunition, thus became much more important. As a result, the tank and the IFV's fortunes became linked.

Even though the XM1 had been designed to cost, that cost was a high one: 710,000 FY 1976 dollars each, over twice the cost of the M60 it was intended to replace.[28] This figure—to be multipled by almost ten thousand, after all—together with the novel design of the new tank, the Army's conspicuous failures in developing technologically advanced weapons over the previous decade, the general suspicion of government, and in particular military activities in the wake of Vietnam and Watergate made the XM1 the object of intense Congressional and media scrutiny.

The GAO presented its findings in a comptroller general's report dated 12 December 1977. It was notably critical of the XM2/3 design. The GAO report asserted that the 1973 Arab-Israeli War had shown the necessity for a properly constituted combined arms team on the modern battlefield. This meant a well-equipped mechanized infantry force. At the same time, the report raised the question of whether the XM2 was in fact the proper equipment.[29] In its investigation, the GAO compared the Army's stated requirements for the tank-IFV team with the actual performance achieved. The civilian investigators pointed out that institutional considerations made coordination of the two vehicle programs difficult:

> To effectively act as a team, the XM-1 and IFV must complement each other. Although the XM-1 and the IFV are to work closely together, two different Army branches are responsible for developing them. The XM-1 tank is under the Army armor branch, which is dedicated to maintaining and operating tanks. The IFV is under the Army infantry branch, which in the past has gone into battle on foot.[30]

GAO thus set out "to determine whether separate development affected the compatibility of the XM-1/IFV as a team and whether recent events affected the Army's. . . justifications for these systems. . . . "[31] Without stopping to define "compatibility"—an omission common to Army sources as well—the GAO compared the two vehicles and concluded that the IFV was significantly inferior to the XM1 in cross-country mobility and slower in acceleration. Furthermore, it presented a larger silhouette with less armor and more pronounced noise and smoke signatures compared to the turbine-powered XM1.[32] Compounding these problems was an absence of doctrinal publications on tank/IFV teamwork. Although TRADOC had produced a groundbreaking series of field manuals on tank/mechanized infantry tactics, the FM 71-series, these dealt with the M60/M113 team, with a different set of characteristics. The GAO report commented:

> In the case of the IFV, the Army is following a concept of developing a new weapons system and then determining how it will be used. Army officials said that the tactics for the IFV will be refined as soldiers train with the vehicle. In fact, this was one reason the Army wanted to field an interim version of the IFV[33]—a plan it has now abandoned because testing did not substantiate its practicality. We believe that the Army should develop its written plan now, so that it may make vehicle modifications, if warranted, in a timely and less expensive manner.[34]

The GAO's final area of concern was the impact of the complexity of the IFV. Aside from concerns about the potential maintenence burden and immaturity of the turret and firing port weapons of the XM2, which were still in development, the investigators were concerned about the role of the IFV gunner, who was responsible for firing and maintaining three different systems: the 25mm cannon, the TOW missile, and the 7.62mm coaxial machine gun; acquiring and maintaining the necessary set of skills was seen as a potentially critical problem.[35] It must be pointed out that the investigators did not take into account the relative simplicity of operation, if not of maintenance, of the TOW system: Israeli gunners had reportedly been brought to reasonable proficiency with four hours' practice under emergency conditions during the 1973 war,[36] whereas Egyptian Sagger operators had to train on simulators even at the height of the war, firing 20 to 30 simulated rounds per day simply to maintain proficiency.[37]

Automotive complexity also concerned the authors of the comptroller general's report. The problematic hydromechanical transmission attracted attention as a major source of mechanical breakdowns. So did the suspension system. These, however, were recognized as developing items, with the potential for improvement. A more fundamental criticism was aimed at the swim-barrier flotation system. The GAO pointed out that the swimming requirement—certainly not "compatible" with the mobility characteristics of the 58-ton XM1, and not found in the *Marder* (see above)—was essentially a badly thought-out aping of the BMP. Swimmability placed a fundamental limit on the XM2's armor protection and added complexity and expense to the vehicle for marginal return, especially given that the Army estimated that only 3 percent of European waterways required a swim capability to cross.[38]

Despite this plethora of complaints, it is worth noting that the GAO's technical criticisms were not singlemindedly against innovative technology. The report, to cite one example, looked askance at the army's decision to omit collective NBC protection from both the XM1 and the XM2/3. In particular, it cited the advances in Soviet offensive chemical capabilities, together with the effort to install such protection on their tanks and BMPs. The latter, it will be recalled, had been introduced explicitly for a projected nuclear battlefield.

The report concluded by pointing out that a reassessment of the IFV was overdue:

> The latest economic analysis of the IFV dates from April 1970. Since then there have been several events which could affect the analysis—enemy threat assumptions have changed, the XM-1 program has replaced the MBT70/XM803 program, the 25-mm gun has replaced the 20-mm gun, the TOW missile launcher has been added, the current design calls for a two-man turret, and IFV costs have increased from an estimated \$99,300 to an estimated \$338,000. Thus an updated analysis of the IFV is long overdue. The Army target date for an updated analysis giving consideration to such changes is January 1978.[39]
>
> The new analysis will compare the reconfigured version of the IFV to (1) an IFV with a one-man turret 20-mm gun, (2) an IFV with a two-man turret 25-mm gun but no TOW, (3) the M113A1, and (4) the Dutch armored infantry fighting vehicle.[40] The new IFV analysis will not consider the possibility of an improved BMP being fielded by the

Soviets. U.S. intelligence sources believe that if the Soviets consider the 25-mm gun a serious threat, they will field an improved BMP in the early 1980s.[41]

In its final recommendations, the comptroller general's report suggested

. . . that the Secretary of Defense determine:

—Whether, considering its advanced state of development, the indicated design changes to make the IFV effective can still be practically made at an acceptable cost.

—Whether a tactical doctrine can be developed that can accommodate the incompatibilities between the IFV and the XM-1 and still provide the effectiveness anticipated from both.

We recommend that, if the design changes and tactical doctrine are unattainable, the Secretary direct the Army to find alternatives to the IFV.

We also recommend that the Secretary require the Army to rejustify using individual protective masks and clothing against chemical warfare rather than equipping its new vehicles with protective systems. [42]

Acting largely on this evaluation, the Office of Management and Budget deleted M2/3 funding from the Carter administration's FY 79 budget. A political struggle then ensued to reinstate the program. In February 1979, at hearings before the House Armed Services Commitee on the Fiscal Year 1979 defense appropriation, Brigadier General Stan Sheridan, program manager for fighting vehicle systems, defended his program, stating that:

The Army feels rather strongly there is an urgent requirement for Infantry and Cavalry Fighting Vehicles to fight side by side with the XM1 Tank. It is the Army's position that the current concept and design of that vehicle . . . meets the requirement and provides improved and versatile firepower; mobility compatible with XM1; sufficiently increased protection to allow infantry and cavalry to fight from within the vehicle; and simplified maintenance.[43]

Under questioning from the committee, Sheridan addressed the GAO report, repeatedly emphasizing the advantages of the IFV's enhanced firepower, particularly with respect to the GAO's concern for the increased complexity of the new vehicle. That complexity, he claimed, was the price of increased capability. He emphasized the considerable developmental progress made since the GAO investigators had done their fieldwork in late 1976 and early 1977, when efforts were still concentrated on the XM723. Sheridan stressed this aspect especially as regarded the GAO's criticisms of the IFV's mobility differences with respect to the XM1.

On the matter of the IFV's survivability and cost-effectiveness, Sheridan pointed out that the two were linked and that the IFV's armor system protected it against 92 percent of weapons in the Soviet motor rifle division, a figure frequently cited afterward, with slight variations, and never, ever, adequately documented.

Toward the end of his testimony, Brigadier General Sheridan expressed his frustration with the stop-and-start course of the program:

I don't mean to be facetious but it seems like about every few years we take the plant called IFV, we pull it out, and look at the roots to see if they are healthy . . . . Let me be totally candid. We are not that far behind today where we would be had we delivered that first vehicle in 1967 which was a little bit better than the 113, and probably equivalent to a BMP. We would have evolved from that day to this probably where we are today. The difference would be that we would have fighting vehicles in our units opposed [*sic*] to the M113. From a development standpoint we are probably where we would have been anyway. We just don't have the hardware.[44]

In mid-April 1978, the Crizer task force reported to Congress, confirming both the requirement for and design of the IFV and CFV, reporting that a vehicle with increased survivability would involve high investment costs, unacceptable delays in fielding the vehicle, and medium to high technical risk.[45]

While the Crizer task force was completing its investigations for Congress, another one was gearing up to answer the concerns of the Office of the Secretary of Defense (OSD) concern about the program. The IFV/CFV Special Study Group formed at Fort Leavenworth, Kansas, in March 1978, under Brigadier General Fred Mehaffey. The Mehaffey group, in conjuction with the TACOM program manager for fighting vehicle systems, was directed to determine the cost-effectiveness of an IFV-based successor for the M113-based Improved TOW Vehicle, and to determine the cost-effectiveness of the IFV and CFV programs against M113 derivatives.[46] The Mehaffey group's recommendations were in the same vein as those of the Crizer task force. It found that the ITV was more cost-effective than the proposed IFV-based TOW vehicle[47] and that M113 derivatives would require extensive reengineering to be even marginally satisfactory, requiring extensive modifications to accept a TBAT-II turret or analogous weapons station, along with a new power plant and suspension. To this end, the Mehaffey group recommended that the Army proceed as planned with the IFV and CFV.[48]

Despite fiscal and political uncertainty, development continued on the XM2 and 3. At about the same time that the Crizer task force reported, FMC delivered the first two prototype TBAT-II turrets, which were fitted to suitably modified XM723 hulls for testing, especially with respect to TOW missile integration. The following month the initial stages of production planning began, as did hot-cell environmental testing of the production configured power train, the Cummins diesel engine being uprated to 500hp over the XM723's original 450 to cope with the increased weight of the XM2.[49] By June, the first TOWs were fired at Fort Irwin, California; thirteen shots out of sixteen were successful.[50]

Troop training for the first phase of operational testing began in August at Fort Carson, Colorado. Curiously, because the final selection for the IFV/CFV main armament had not yet been made, the troops involved were trained on both of the final candidates, the self-powered Ford Aerospace/Oerlikon weapon and the externally powered Hughes design, along with the 7.62mm M240C coaxial machine gun. This weapon was also just entering U.S. service as secondary armament for the XM1 and XM2/3, after almost two decades service elsewhere as the FN MAG 58 general-purpose machine gun.[51]

In October 1978, the deleted procurement funds were reinstated by Congress. The previous public-law mandate for production delivery to commence by May 1981 was retained. In January 1979, the Hughes externally powered 25mm cannon was at last selected as the main armament, a month after the first of the eight prototypes was delivered for testing. Operational testing, intended to determine how well the vehicle met user expectations, and developmental testing, to establish its conformity with the formal requirements, occupied most of 1979. These tests, as intended, revealed a sizeable number of shortcomings, particularly in the TOW launcher and Integrated Sight Unit. These problems yielded to fixes ranging in scope from the application of a common adhesive to a motor connector all the way to extensive system redesigns.[52] These remedies, however, cost time and money, which alarmed Congress. The unit cost of the IFV, in constant FY 78 dollars, went from $370,000 to $472,000,[53] a cost which was quickly revised upwards, to $495,000[54] due to tooling problems at FMC. Taking the inflation rates of the era into account, this made the rollaway cost (i.e., the cost of a completed, fully equipped, M2 or M3) $557,000 in FY 80 dollars, averaged over a projected purchase of seven thousand units.[55] Army representatives maintained that the IFV was still highly cost-effective, however, claiming that an IFV/CFV-equipped force, while 5–12 percent more expensive than one equipped with the M113A2, was 80–120 percent more effective. As with the claim that the IFV was resistant to 92 percent of Soviet weapons (see above), no elaboration for this claim was given.[56]

Meanwhile, the XM2/3 passed the Army Systems Acquisition Review Council Milestone III review, and were type-classified Standard on 20 December 1979, along with the M242 25mm cannon and the M231 firing port weapon. The Department of Defense concurred on 22 January 1980,[57] resulting in final approval for production by the Secretary of Defense on 1 February. Testing of the eight pilot vehicles continued apace; development testing was conducted through June, followed by user testing at the Armor center, Fort Knox, Kentucky, from June to August.[58]

FMC delivered the first production vehicle by the May 1981 deadline, and as stocks built up for issue, on 20 October 1981, the M2 IFV was officially dedicated in honor of General Omar Bradley in a ceremony at Fort Myer.[59]

In short, then, the newly expanded MICV program set about redesigning its vehicle to accommodate the new mission. Merger of the two disparate and inchoate sets of requirements led to delays and cost escalation, as did the accommodation of a doctrinally mandated antitank capability. This expense and delay, together with the design trade-offs necessary to accommodate all the above, made the program controversial, leading to its near termination. Nonetheless, the newly named Bradley Fighting Vehicle entered production in

1981, and unit service in 1983. At last, the Army had an IFV; it was now necessary to decide what to do with it.

NOTES

1. "Tank-Automotive News," *National Defense* 71 (September 1987): 25.

2. That is, the control set was equipped with a tracking system that piloted the missile automatically while the operator simply kept his sight trained on the target. Earlier missiles, such as Sagger, required the operator to fly the missile to the target by hand with a joystick. Note that "automatic" is a shorthand used here for simplicity; the full name of the technique is "semi-automatic command to line-of-sight," or SACLOS.

3. Seward Smith, *Human Factors Implications of MICV OT II for Infantry Fighting Vehicle (IFV) Development* (Alexandria, VA, Department of the Army, Office of the Deputy Chief of Staff for Personnel, U.S. Army Research Institute for the Social and Behavioral Sciences, 1968), 9.

4. Joseph B. Hayes, "MICV plus TOW," *National Defense* 58 (March–April 1975), 358.

5. Ulrich Kleeman, "*Streiflichter zur Kriegsfuehrung in Nordafrika*," Foreign Military Studies Manuscript D104, pp. 4–8, quoted in John A. English and Bruce I. Gudmundsson, *On Infantry*, rev. ed. (Westport, CT: Praeger, 1994), 107.

6. The Dragon used a novel propulsion system based on multiple paired thrusters which simultaneously provided lift, forward propulsion and directional control during flight by the timing of their firing as the missile's airframe rolled. This arrangement was combined with an unstable bipod mounting which required the gunner to stay absolutely still in a kneeling or seated position for the entire engagement—approximately ten seconds at maximum range—for the automatic guidance system to work. Doing so was difficult even on the range.

7. That is, from above the troop compartment; note that the date precedes the Larkin report.

8. DePuy to General Fred Weyand, 29 April 1975, in Paul H. Herbert, *Deciding What Needs to Be Done: General William DePuy and the 1976 Edition of FM 100-5, Operations* (Ft. Leavenworth, KS: Combat Studies Institute, U.S. Army Command and General Staff College, 1988), 78.

9. U.S. Congress. Senate. Committee on Armed Services. *Fiscal Year 1978 Authorization for Military Procurement, Research, and Development, and Active Duty, Selected Reserve, and Civilian Personnel Strengths: Hearings before the Committee on Armed Services, United States Senate, on S. 1210.* 95th Congress, 1st Session, 1977, 5928. Hereafter abbreviated "SASC 1978."

10. "Developing Army's Bradley: The Wages of Compromise," *Army* 37 (July 1987): 71.

11. While commandant of the NATO Defense College, Rome, *Bundeswehr* Lieutenant General Franz Uhle-Wettler pointed out that a specialized antitank self-defense missile would be much more useful on an IFV than the preexisting medium- and long-range missiles fitted; unfortunately, no such weapon (a short-range missile with a quick reaction time) exists despite the acquisition resources put into IFVs by major armies (Franz Uhle-Wettler, "Infantry versus Tank: Common Sense and High Technology," *NATO's Sixteen Nations* [May–June 1984]: 53).

12. United States. Department of the Army. Tank-Automotive Command. Program Manager, Mechanized Infantry Combat Vehicles. "MICV Systems Messages." United States Army Tank-Automotive Command Archives, Warren MI , "MICV Systems Message #3," Inclosure (*sic*) 1.

13. Michael S. Hackney, "Understrength Platoon," *Infantry* 72 (November–December 1982): 10. Lieutenant Hackney was a mechanized infantry platoon leader ca. 1980.

14. United States. Department of the Army. Tank-Automotive Command. Program Manager, Mechanized Infantry Combat Vehicles. "MICV Systems Messages." United States Army Tank-Automotive Command Archives, Warren MI, "PM Message 6," 5ff. These messages wil hereafter be cited as "PM Message [number]."

15. Richard M. Ogorkiewicz, *Technology of Tanks* (Coulsdon, Surrey, UK: Jane's, 1991), 303.

16. Richard M. Ogorkiewicz, "Infantry Armored Vehicles: Origins and Future Trends," *International Defense Review* (May 1990): 512. One of the most thorough published rationales for the heavy infantry fighting vehicle may be found in Richard Simpkin, *Mechanized Infantry* (London: Brassey's, 1980), chapter 5. Grossly compressed, Simpkin's thesis was that the main battle tank had (and has) grown so specialized as an antitank system itself and that the threat from other antitank weapons had grown so, that there were many tactical situations in which IFVs, with their handier armament and dismountable troops, would have to lead, and that in any event, manpower and morale considerations in the NATO scenario called for higher levels of protection. Simpkin countered the cost and weight objections normally leveled at this approach by showing that for large fleet sizes, commonality between MBT and IFV components could realize economies of scale, and that the potential for hiving off secondary functions from the MBT to a complementary IFV, capable of closer cooperation than the lighter vehicles fielded so far, could drive weight down. In this book, as in several of his others, Simpkin emphasized that for the NATO powers, the critical resource was manpower rather than money or materiel. Certainly the United States was a prime historical example of this thesis; the nation had run perilously short of riflemen in its last three wars, but never short of funds.

17. SASC 1978, 5916.

18. James Burton, *The Pentagon Wars: Reformers Challenge the Old Guard* (Annapolis: Naval Institute Press, 1993), 53.

19. John L. Romjue, *From Active Defense to AirLand Battle: The Development of Army Doctrine, 1973–1982* (Ft. Monroe, VA: U.S. Army Training and Doctrine Command, 1984), 13; see William S. Lind, "FM 100-5, Operations: Some Doctrinal Questions for the United States Army," *Military Review* (March 1977): 54–65, for Lind's extended critique of the Active Defense doctrine.

20. SASC 1978, 5918–23.

21. SASC 1978, 5923.

22. SASC 1978, 5924.

23. That is, copy using U.S. components and technologies.

24. SASC 1978, 5927. Note that the Dutch vehicle alluded to here is the YPR-765, a licence-built version of the private-venture AIFV FMC had derived from the M113 (see chapter 5).

25. Simpkin, *Mechanized Infantry*, 29.

26. "PM Message 7," 1.

27. This excursus is based on the comprehensive technical survey in R. P. Hunnicutt, *Abrams: A History of the American Main Battle Tank, Volume II* (Novato, CA: Presidio, 1990), part III.

28. U.S. Congress. General Accounting Office. Comptroller General. *The Army's Proposed Close Combat Vehicle Team* (Washington, DC: GAO, 1977), 1; hereafter cited as "GAO 1977."

29. GAO 1977, i.

30. GAO 1977, 1f.

31. GAO 1977, 3.

32. GAO 1977, 4.

33. The XM723 MICV with M139 20mm cannon.

34. GAO 1977 11. Cf. I. B. Holley, Jr., *Ideas and Weapons* (New Haven, CT: Yale, 1953); the tendency for U.S. Army procurement bureaucracies to make doctrine fit its desired weapons dates back at least as far as World War I.

35. GAO 1977, 15ff.

36. O'Ballance, *No Victor, No Vanquished*, 162.

37. Isby, *Weapons and Tactics*, 145f.

38. GAO 1977, 20.

39. The Crizer task force commissioned in October 1977.

40. A 13,000-pound, ten-man, 25mm-armed vehicle derived from FMC's private-venture M113A1PI/XM765 design (see chapter 5).

41. GAO 1977, 25. The Soviets did so, in fact: the BMP-2 carried a 30mm automatic cannon and an improved semiautomatic command to line-of-sight missile in a two-man turret, with a resultant loss of two men from the vehicle's dismount strength. Cf. the 1976 metamorphosis of the XM723 to the XM2.

42. GAO 1977, 27.

43. U.S. Congress. House. Committe on Armed Services. *Hearings on Military Posture and H.R. 10929: Department of Defense Authorization for Appropriations for Fiscal Year 1979*. H.R. 10929, 95th Congress, 2nd session, 1978, 171. Hereafter cited as HASC 1979, 171.

44. HASC 1979, 209f.

45. "PM Message 7," 1.

46. Kaufman, "Organizations, Technology, and Weapons Acquisition," 409.

47. This was a considerable irony given the Army's rejection of the XM765 and other M113 product improvements in the 1960s due to the cost impact of a two-vehicle fleet for the infantry (see chapter 5).

48. Kaufman, "Organizations, Technology, and Weapons Acquisition," 414.

49. "PM Message 7," 3.

50. "PM Message 8", 2f.

51. *Ibid.*; "MAG" stood for *Mitrailleuse à gaz,* "gas-operated machine gun."

52. U.S. Congress. House. Committee on Appropriations. Subcommittee on the Department of Defense. *Department of Defense Appropriations for 1980*, part 7. 96th Cong., 1st sess., 1979, 132–36; hereafter cited as Appropriations 1980.

53. *Ibid.,* 109.

54. U.S. Congress. House. Committee on Armed Services. *Hearings on Military Posture and H.R. 6495 [H.R. 6974]: Department of Defense Authorization for Appropriations for Fiscal Year 1981* (96th Cong. 2nd sess., 1980), 857; hereafter cited as HASC 1981.

55. U.S. Congress. Senate. Committee on Armed Services. *Department of Defense Authorization for Appropriations for Fiscal Year 1981* (S.2294, 96th Cong. 2nd sess., 1980), 2333; hereafter cited as SASC 1981.

56. SASC 1981, 2335.

57. HASC 1981, 861.

58. United States. Congress. House. Committee on Armed Services. *Hearings on Military Posture and H.R. 2970 [H.R. 3519]: Department of Defense Authorization for Appropriations for Fiscal Year 1982*, part 2 (97th Cong. 1st sess.), 352.

59. TACOM AHR 1983, 225.

M113A1 armored personnel carrier. (Photo courtesy of United Defense)

XM765, an attempt to improve the M113's combat capabilities at minimum cost. (Photo courtesy of United Defense)

XM723 MICV. (Photo courtesy of United Defense)

M113A1PI, a further development of XM765, enhanced with armor technology developed for the XM723 MICV. A turreted version, the Armored Infantry Fighting Vehicle has been adopted by the armies of several nations. (Photo courtesy of United Defense)

M114A1 armored command and reconnaissance vehicle. (Photo courtesy of the United States Army)

M551 Sheridan armored reconnaissance/airborne assault vehicle. (Photo courtesy of the United States Army)

XM800 Armored Reconnaissance Scout Vehicle. (Photo courtesy of United Defense)

The Bradley: original version, in this case an M2 IFV. (Photo courtesy of the United States Army, apparently originating with FMC Corporation, a predecessor of United Defense)

The Bradley: A2 version with enhanced armor. (Photo courtesy of United Defense)

The Bradley: A3 version with enhanced sensors and electronics, notably Commander's Independent Thermal Viewer on turret roof opposite TOW launcher and Gunner's Primary Sight. (Photo courtesy of United Defense)

Chapter 8

Adoption and Adaptation: The Bradley in Service, 1983–1988

> [An IFV] squad leader would have to be capable of replacing both an M60 tank commander and a straight infantry squad leader. He would also be required to perform as a combined arms platoon leader on a small scale. In sum, his daily tasks would be virtually impossible.
>
> —Lieutenant Colonel Ralph Hallenbeck, mechanized infantry battalion commander[1]

What manner of vehicle had the Army ended up with after such a prolonged development? The newly christened Bradley Fighting Vehicle, weighing 24.5 tons combat loaded, retained the basic hull configuration of the XM723 MICV. The main part of the hull was fabricated from the same 5083 aluminum alloy as the M113, although the plating was thicker on the M2. A different alloy, 7039 aluminum—harder to work, but more ballistically resistant—was used on the turret front and the upper hull sides; both were sloped to increase the effective protection level, both by presenting a greater apparent thickness to projectiles normal to the hull, and by increasing the probability that they would bounce off at extreme angles of attack. Additional side protection was provided by a spaced laminate of high-hardness steel and aluminum developed by the Army's Ballistics Research Laboratory. This armor system was designed to provide a form of defense in depth: the outer steel layer was intended to disrupt incoming projectiles, causing them to deform and tumble so that the softer aluminum layer inside could then stop them more easily. The protective system was completed by a steel antimine appliqué (omitted in the M3 CFV) covering the bottom front of the vehicle. The whole was reckoned to provide overhead protection from 155mm airbursts and protection from direct hits by 14.5mm heavy machine gun fire (i.e., from the main armament of Soviet BTR-series armored personnel carriers) all around.

The transmission and the engine, a 500hp Cummins VTA-903 diesel capable of driving the vehicle at 41mph on the road, were mounted forward and to the right. Two tanks amidships behind the turret held 175 gallons of fuel. This supply was nominally capable of taking the Bradley three hundred miles, a range matching that of the M1 tank. To the left of the power pack sat the driver, beneath a novel clamshell-type hatch which could be opened halfway to provide direct vision while maintaining overhead cover. As an alternative, the hatch could be closed for complete protection, vision being provided by periscopes. One periscope could be replaced by an AN/VVS-2 passive light-amplifying device for night driving.

As with the XM723, the turret was mounted amidships and slightly off-center to the right. Although in the same location as before, the new two-man turret was a larger and more elaborate structure. It embodied as many functions, if not more, than the relatively spacious three-man turret of the main battle tank. The vehicle commander and gunner sat to the left and right, respectively, behind the main gun. This weapon, the 25mm M242 Chain Gun™ manufactured by Hughes Helicopter, was chosen in part because its design minimized intrusion of the gun's receiver into the crew's space. In addition, the weapon's action promised higher reliability than traditional self-powered automatic weapons, being externally powered. Rather than using the energy of the gas or recoil of the fired ammunition to work its action, the Chain Gun employed a loop of double-row roller chain driven by a 1.5hp electric motor at rates of 100 or 200 rounds per minute; it could also fire single shots. The use of external power made for a simpler mechanism and minimized the potential for malfunctions due to ammunition failure.

The M242 was supplied with 300 ready rounds of ammunition below the gun mount. This supply was of two types, selectable with the throw of a switch: 225 rounds of high explosive (HE) for use against unarmored targets, and 75 rounds of Armor-Piercing Discarding Sabot (APDS) ammunition. The armor-piercing round was nominally capable of penetrating 66mm of homogeneous steel armor at 0 degrees obliquity at 1000 meters; it was thus reckoned easily capable of defeating the BMP, with its 14mm turret armor sloped at 45 degrees.[2] Alongside the main weapon was mounted a coaxial M240 7.62mm machine gun, fed from an 800-round box to the right of the commander. Both weapons were separated from the crew by a gas-tight door preventing fumes from entering the fighting compartment. Both weapons' spent cartridge cases were ejected overboard. Nor were all the turret weapons inside. Two 4-barrelled smoke grenade launchers were mounted at the front corners to provide quick screening, while a two-tube launcher for TOW antitank missiles was folded along the left side of the turret. When raised, it allowed the missiles to be fired from partially defiladed positions. This feature helped lessen the vulnerability of the Bradley when launching the wire-guided TOW, which could not be fired on the move.

The TOW, along with the rest of the Bradley's turret weapons, was aimed using the Integrated Sight Unit (ISU). This device provided both visible light and imaging infrared channels at four- and twelve-power magnification. To allow the TOW launcher to be used and observation to be conducted from behind terrain features, the ISU's head protruded above the turret roof. The ISU was comparable to the XM1 tank's primary sight, although the Bradley's sight lacked the latter's full-solution computerized fire control, as well as its associated laser rangefinder.

The Bradley commander could see through the ISU by means of an optical relay. He was also provided with two radios, an AN/GRC-46 and an AN/GRC-160. The latter could be dismounted and man-packed for ground use.

The remainder of the vehicle was given over to the dismountable element of the squad. Two men were seated behind the driver to the left of the turret basket, the front one facing the rear and the rear one frontward. Separated from these by a stowage area for five antitank missiles, a third crewman sat in the left rear corner with his back to the side of the hull. Two more men in the rear were seated facing the ramp. Finally, a sixth man was seated facing forward at the right rear of the turret basket. Each seat's occupant was provided with a fixed periscope and firing port; rather than using individual weapons through them, in the fashion of other mechanized infantry vehicles, six of the specialized M231 Firing Port Weapons (FPW) developed earlier for the XM723 MICV were provided. The FPWs had no usable sights, so fire control was achieved through the use of tracer ammunition.

In addition to the dismount team and its supplies, weapons, and equipment, the rear compartment of the BFV stowed a substantial amount of ammunition: 600 additional rounds of 25mm, 1,540 of 7.62mm for the coaxial machine gun, 4,200 rounds of 5.56mm ammunition for the Firing Port Weapons, and up to five additional TOW rounds, exchangeable for the dismount team's lighter, man-portable, Dragon weapons. The dismount team itself disposed of more weapons than it had members. In addition to their M16A1 rifles, with another 2,520 rounds of 5.56mm ammunition, they were equipped with an M60 machine gun (with another 2,200 rounds of 7.62mm), two 5.56mm M249 Squad Automatic Weapons, an M203 40mm grenade launcher, a tracker for guiding Dragon antitank missiles, three M72 unguided light antitank rockets, and assorted flares, mines, and grenades; in essence, this was the armament of the eleven-man M113-equipped squad, concentrated into the smaller Bradley-mounted group.

Externally, the M3 Cavalry Fighting Vehicle was almost indistinguishable from the M2. Internally, however, the M3 traded the dismount team and their gear for two scouts seated at the extreme rear of the troop compartment, and a larger ammunition supply for the vehicle's weapons: 1,200 stowed rounds of 25mm, 3,600 stowed rounds of coaxial machine gun ammunition plus another 5,060 for an additional M60, and ten TOW reloads. The FPWs were omitted from the M3, and their mountings plated over. Originally, the M3 was intended to carry a motorcycle and a ground surveillance radar set to enhance its reconnaissance capabilities. These items, however, were dropped during development because of expense, stowage space, and, in the case of the gasoline-fueled motorcycle, safety.

Given this extensive panoply, so much more complex and powerful than that of the predecessor mechanized infantry squad, the question of how it should be used naturally arose. As Brigadier General Sheridan had pointed out (however obliquely) in his congressional testimony rebutting the GAO's criticism of the XM2/3 program, TRADOC had conducted a comprehensive 1977 revision of tactical field manuals to reflect the Active Defense doctrine. This effort had laid some doctrinal groundwork for the Bradley's employment, especially the groundbreaking FM 71-series dealing with tank-mechanized infantry teams of company to brigade size.

Completely new equipment at the lowest levels, however, required new publications. To this end, as early as 1980, the U.S. Army Infantry School at Fort Benning, Georgia, had produced a provisional manual for the IFV-mounted mechanized infantry squad based on operational testing. The resulting document, *Special Text 7-7-1 FY80, The Mechanized Infantry Platoon and Squad (IFV)*, is a window on the Infantry branch's attitude toward the new weapon. It is hard to describe that attitude without resorting to use of the term "schizophrenic." *ST 7-7-1* begins by attempting to describe the battlefield role of IFV-equipped mechanized infantry:

> The M-2 Infantry Fighting Vehicle (IFV) represents the most dramatic improvement in infantry combat capability since the introduction of the machinegun. The infantry will now have unprecedented firepower, armor protection, and battlefield agility while still retaining its traditional capability of fighting on foot.
>
> The IFV is not an improved Armored Personnel Carrier (APC); it is truly a fighting vehicle. This is a new dimension infantrymen must master. The fundamentals of current tactical doctrine remain essentially unchanged. They must, however, be modified to capitalize on the IFV's capabilities.
>
> The more conservative thinkers will tend to regard the IFV as an improved APC or "battle taxi." The other extreme will think of the IFV as a light tank. The correct role of the IFV is in between these two extremes, but probably slightly more toward the light tank.[3]

Having established this fairly radical position, the author of the introduction then sensibly backs away from it:[4]

> There will be occasions when the infantrymen can observe the battlefield and fire their weapons under the protection afforded by the IFV's armor without dismounting, but infantrymen must not regard the IFV as invulnerable. The armor does provide excellent protection against small arms fire and shell fragments, but it does not protect against tank or antitank guns, missiles, or rockets.
>
> This means that the infantry leader must carefully analyze every situation and weigh the advantages and disadvantages of mounted and dismounted combat. Because of the

increased capability of the IFV, mounted combat will more often be the preferred method than with the APC.

Many combat tasks can only be performed dismounted. This is why there are infantrymen in the vehicle. Now, however, when infantrymen dismount to perform their traditional tasks they will have unprecedented supporting firepower from the IFV itself.

In the past, mechanized infantry used the APC as a means of battlefield transportation. Once at the scene of a fight the dismounted infantry became the center of interest and the APC merely added to the squad's firepower with its caliber .50 machinegun.

With the IFV this changes. The dismounted infantry is still critically important but the IFV is of equal importance and in some situations even more important. Hence the vehicle and the dismounted infantry together become the center of combat operations. Both offensive and defensive (mounted and dismounted) tactics must be built around the IFV. The infantry must grasp the fact that the squad is a part of the IFV and the IFV is a part of the squad. THE TWO ARE ONE SYSTEM.[5]

Significantly, however, the body of ST 7-7-1 drew heavily on its predecessor, the 1977 FM 7-7, even to recycling graphics and appendices. As with earlier practice, the platoon was organized into a headquarters and three squads, each in their own IFV. The headquarters consisted of a platoon leader, platoon sergeant (traded to another IFV for a rifleman so as not to risk the entire headquarters in one vehicle), assistant platoon sergeant, radio-telephone operator, and attached medic and forward observer team, plus driver and gunner. Each squad was organized into an IFV team and a dismount team. The IFV element of the platoon was seen as providing a base of fire for dismount team maneuver; the M2 was, in short, regarded by the Infantry School as very much an *infantry* fighting vehicle, its 25mm cannon serving as an enormous analog of the Browning Automatic Rifle in a World War II rifle squad.

Despite the exceptionalism claimed in ST 7-7-1's introduction, then, the new capabilities of the Bradley were in the main treated as quantitative enhancements rather than revolutionary ones. Little emphasis, for instance, was placed on the exploitation of the stabilized main armament to enhance mounted movement or to suppress antiarmor threats in support of tanks. Similarly, the prospect of "light tank"–like independent maneuver by the IFV was minimized. To some extent, the relative conservatism of the new manual was defensible; the 1977 edition of FM 7-7 placed considerable stress on mounted operations, using the weapons organic to the squad and carrier from the M113's open hatches. In essence, then, the Infantry branch went, with minimum expenditure of effort, from propounding perhaps the most aggressive of APC doctrines to perhaps the most conservative with respect to IFVs.

Doctrinal conservatism, however, largely stopped at the squad and platoon level. From its inception, the Active Defense concept, meshed as it was with the political and geographic conditions of Central Europe, had come under criticism. Its orientation toward the defense, its emphasis on winning initial engagements of forecast conflicts to the extent of deemphasizing the provision of designated reserves, and the attrition-intensive view of operations all clashed with traditional military imperatives (see chapter 7).[6] As a result, almost from the publication of the 1977 edition of *FM 100-5, Operations*, TRADOC, now commanded by General Donn A. Starry, set about revising Active Defense to emphasize aggressive, maneuver-based tactics and attitudes. The resulting doctrine, dubbed AirLand Battle, was expounded in a new edition of *FM 100-5*, issued in 1982. AirLand Battle foresaw an expanded battlefield involving deep operations by maneuvering armored forces with extensive rotary and fixed-wing air support. It aimed to thwart Soviet-style massed armored movements, especially second- and subsequent-echelon attacks, while defeating the first echelon in detail.[7]

Along with new doctrine came new organization. The 1977 doctrinal publications issued under DePuy had been attuned to the ROAD organizations that had been in place for a decade and a half. After TRADOC issued the Active Defense doctrine, however, the command spent a year, from September 1978 to October 1979, studying a new divisional organization to accommodate both the new doctrine and the massive reequipment scheduled for the 1980s to deal with the bloc obsolescence brought on by the Army's inability to modernize during the Vietnam War. This Divisional Restructuring Study (DRS) led to a thoroughgoing reorganization of U.S. Army divisions.

The "Division 86" initiative began by partially abandoning the ROAD concept of a common base for all divisions. The standardized divisional base was a victim of a double bind: the perceived increase in the combat power of Soviet-style forces called for enhanced firepower from increasingly heavy equipment such as the BFV and the Abrams tank; supporting these called for a large divisional base. The prospect of projecting forces potentially anywhere in the world, on the other hand, militated in just the opposite direction, placing a premium on strategic mobility. To this end, two basic divisional organizations were promulgated.

One was a light infantry division (LID) with a strength of approximately 10,000, essentially supplanting the old ROAD infantry division with its small armor component and extensive support units. The LID's major constraint in the force design process was that it was to be deployable in 500 C-141 airlift sorties, a fraction of the number required for an armored or mechanized division. As a result, it totally lacked armored vehicles. Other heavy equipment was provided sparingly, as was organic transport.

The other organization was a heavy division almost twice that size, to replace the ROAD mechanized infantry and armored divisions. This type of formation was, if anything, even more standardized than its predecessors. The two division types differed only in their nominal complement of line battalions. Mechanized infantry divisions nominally had four tank and six mechanized infantry battalions; armored divisions had four mechanized and six tank. The ROAD concept of the three maneuver brigade headquarters controlling the battalions on an ad hoc basis, inherited in turn from the armored division stucture evolved during World War II, was retained, as was the practice of cross-attaching companies between mechanized infantry and armored battallions to form combined arms task forces.

At this point, however, the new organization diverged from the old. The designers of Division 86 intended the main combined arms headquarters to be at brigade and battalion level, with each company operating a single weapons system. It was thus the force designers' intention to abandon the earlier practice of subdividing task forces into company-sized combined-arms. Companies themselves, meanwhile, were made more numerous, more homogeneous, and smaller,[8] and the headquarters companies of armored and mechanized infantry battalions were made as similar as possible. The tank battalion therefore went from three companies of three platoons of five tanks each to four companies of three platoons of four tanks each. Similarly, a mechanized infantry battalion went from three companies of three platoons, each with its own TOW and mortar sections, to four homogeneous line companies of three platoons, with a fifth, specialized antiarmor company. The latter was equipped with twelve M901 Improved TOW Vehicles, M113-derived vehicles with missile launchers capable of being elevated to exploit cover. All the mechanized infantry company mortar sections were abolished; the battalion's mortars, now all heavier 4.2-inch weapons, were concentrated in a battalion mortar platoon, as they were in tank units.

The shift to this new "J-series" organization and the fielding of the Bradley were more or less concurrent. At Fort Hood, Texas, on 28 March 1983, the first mechanized infantry unit to be equipped with the Bradley, the 2d Armored Division's 1st Battalion, 41st Infantry Regiment, commenced its New Equipment Training (NET).[9] Due to the organizational changes involved and the relative complexity of the new vehicles, Bradley NET entailed a great deal more than an exchange of equipment. This part alone was a complex enough task, as 2-41 Infantry's commander pointed out in referring to his unit's transition; each battalion's having to provide facilities for a sixth company and thirty-one more and larger armored vehicles had a serious logistic and administrative impact.[10]

At the same time, however, operating the Bradley required a wider range of skills than were previously cultivated even in the earlier H-series mechanized infantry units equipped with the M113. One index of the extent of the reorientation required may be seen in the reassignment of mechanized infantrymen trained on the Bradley to an entire new mechanized infantry Military Occupational Specialty (MOS), 11M, where the old MOS 11B had previously applied to all infantrymen, mechanized or not.

The new designation was well founded; in the thirty-seven days allotted to NET, quondam 11Bs had to learn maintenance, tactical, and gunnery skills formerly associated with tankers. The training of the vehicle and dismount teams themselves was relatively straightforward except that "a Bradley battalion would add between 24 and 36 days of vehicle and gunnery training to its already full annual program," in the words of 2-41's commander.[11] Coordination between the mounted and dismounted training syllabi, however, was problematic; doctrine demanded the treatment of the Bradley-equipped squad as a single entity, yet the complexity and time demands of training the whole organization exerted considerable pressure to split the squad's training into infantry and gunnery tracks.[12] Some indication of the difficulty of the problem may be seen by noting that *FM 23-1, Bradley Fighting Vehicle Gunnery*, the principal Bradley gunnery manual, approaches FM 7-7J, the principal Bradley tactical manual, in bulk. It deals, after all, with the maintenance and operation of four separate types of vehicle-mounted weapons.

The problems of accommodating all these new skills fell hardest on squad leaders and their assistants. Doctrine required that both the squad leader and assistant squad leader be capable of commanding either the vehicle or the dismount team, effectively requiring both men to combine the expertise of a tank commander (who had to be a fairly competent gunner and driver in his own right) and a rifle squad leader.[13] As one mechanized infantry battalion commander framed the problem:

> Imagine for a moment that each mechanized infantry platoon under the current organization was equipped with four Bradley infantry fighting vehicles (IFVs) instead of M113s. How many squad leaders (sergeants or staff sergeants) could do all of the following tasks simultaneously?
>
> —Supervise the maintenance of an IFV with its stabilized turret, chain gun, and TOW missile systems,
>
> —Train an IFV crew to "tank crew proficiency."
>
> —Train a six-to-eight-man dismounted infantry squad to full proficiency, while also accounting for and maintaining all the squad equipment.
>
> —Employ the IFV and the dismounted element in battle as a coordinated unit, despite the physical separation often required to make the most of the characteristics of each.
>
> Such a squad leader would have to be capable of replacing both an M60 tank commander and a straight infantry squad leader. He would also be required to perform as a combined arms platoon leader on a small scale. In sum, his daily tasks would be virtually impossible.[14]

The trend toward technological sophistication that forced the same group of men to train as both an armored fighting vehicle crew and as an infantry squad, however, also brought some relief to the time and financial problems involved. A key element in the training of Bradley squads, both infantry and cavalry, was the Unit Conduct of Fire Trainer (U-COFT), a computerized simulator—essentially derived from contemporary video game technology—which allowed vehicle teams to train with a fair degree of realism. Nor were solely technological solutions to the problems of training fielded. Integral to the Army-wide Bradley training program was an effort to develop a corps of senior NCOs to serve as master gunners to assist company and battalion commanders with the training problems inherent in the new systems.

With these aids, new equipment training proceeded, as did assimilation of the Bradley into the training base at both the Armor and Infantry Schools. By the end of 1983, 450 Bradleys had been fielded; a year later, 1,100.[15] By then, enough experience with the vehicle had been gained for improvements to be incorporated into a new version. The M2A1 and M3A1 were characterized by relatively minor improvements in human and automotive engineering with respect to the original Bradleys. Most of the changes dealt with the TOW missile system, which was modified to accommodate more capable variants of the antitank missile. The new TOWs were fitted with enlarged and enhanced warheads, capable of defeating the improved armor arrays protecting the new generation of Soviet tanks.[16]

Another significant change was the incorporation of a collective NBC protection system, as originally intended for the first MICV so long ago. Although the new cavalry vehicle's system of filters and ventilators applied to the whole crew, the new model of the IFV, the M2A1, served only the vehicle team. The men of the dismount team, in contrast, continued to rely on their personal MOPP[17] gear. In a rather disingenuous statement, the test director for the Army's Combat Systems Test Activity defended this arrangement as superior to the shirtsleeve environment provided by the Soviet BMP's more elaborate overpressure NBC protection system, saying: "The Soviet soldier must change into his protective clothing before leaving his armored personnel carrier when he's in a chemical, biological, or nuclear-contaminated area. Our troops are able to leave the M2E1 at anytime through use of the MOPP gear they already wear."[18]

In addition to these internal changes, the Bradley Cavalry Fighting Vehicle gained external ones distinct from the IFV version. New CFV's had the firing ports entirely omitted instead of blanked off. They also replaced the original domed cargo hatch, designed to give overhead protection while reloading the TOW launchers. The new hatch was fitted with periscopes to enhance the observation capabilities of the scouts in the rear compartment.

As experience accumulated with the BFV, a number of trends emerged. One was the perception of a growing doctrinal and functional gulf between the light and mechanized infantry communities. The most rigorous analysis of this phenomenon was made by then-Colonel Huba Wass de Czege, an infantry officer and one of the major architects of the AirLand Battle doctrine. Writing in *Infantry*, he pointed out that the infantry role had broadened over time to encompass a wider range of missions and equipment than formerly and that organizations and attitudes had not really kept up. U.S. doctrine recognized two varieties of infantry, mechanized and light (the latter including airborne and airmobile infantry); Wass de Czege made the case that that there were, in fact, three distinct subgroups of infantry, with differing material and training requirements:

Infantry missions cover a wide functional range. Because of this, I believe we need three basic kinds of infantry today.

—We need infantry whose primary mission is to support the advance of the tank. Let's call this *armored* infantry.

—We need infantry whose primary mission is to hold ground and to take fortified or infantry-defended positions. Let's call this *regular* infantry.

—We need infantry that is strategically, operationally, and tactically mobile using Army or Air Force aircraft and that can fight highly mobile tactical engagements in difficult terrain. Let's call this *light* infantry. (Light infantry may have several variants, such as air assault and airborne.)

. . . The important thing here is to resist the trend toward only two kinds of infantry—armored and light. Much of the debate today over how to use the Bradley-equipped infantry and the new light infantry results from trying to use either force as regular infantry. We need to develop a third type of infantry—a regular infantry—for use in that middle range.

We may already have such a force. In fact, we have basically had it all along in our M113-equipped mechanized battalions—especially when they were trained to fight using tactics suited to their equipment. That's worth thinking about![19]

This last caveat can be construed as implicit criticism of TRADOC's effort to fit the square peg of mechanized infantry units still equipped with the M113 into the round hole of mechanized infantry doctrine oriented toward the Bradley. A new version of *FM 7-7*—as opposed to *FM 7-7J*, the platoon- and squad-level publication for Bradley-equipped units—had just been issued in March 1985.[20] This manual emphasized mounted combat even more extensively than the 1977 edition. With the addition to the M113 of the M175 vehicular Dragon mount, there was a tendency on the part of the authors to regard the vehicle as a surrogate IFV rather than a simple transport. In addition, their manpower scales were adjusted to J-series levels, with nine-man rifle squads and centralized heavy weapons. This arrangement made Wass de Czege's category of "regular" infantry missions more difficult to support. As one critic posed the problem:

. . . is anyone at Fort Benning looking at the *operational* employment of M113 units as "regular" infantry along the lines suggested by Colonel Huba Wass de Czege. . . ? Neither Bradley battalions nor those in the new light infantry divisions are really suitable for the hard slugging required to attack well-entrenched enemy or defend a battle position in depth. Bradley battalions, although long on firepower, are short on dismounted infantrymen; light battalions don't have the firepower or the sustainability needed. I fully realize that light infantry prefers to attack the enemy's weak spots, but sometimes you can't find a weak spot to attack.[21]

This opinion, however, was subject to dispute. In particular, it flew in the face of the doctrinal monism promoted within the infantry community ever since the ROAD initiative of the early 1960s (see chapter 3). As voiced by one member, it argued thus:

> I personally believe that there are not three different kinds of infantry and that there is no need for three. There is only one type of infantryman, and he is employed differently in different scenarios and units. . . .
>
> Having served in light infantry, airborne infantry, and mechanized infantry, I see no real differences beyond extra equipment and employment. Despite all the arguments to the contrary, I have found it quite easy to move from one "kind" of infantry to another. The basic training required is the same, and the tactical employment of the different "kinds" is not all that difficult.
>
> The idea of institutionalizing three different types of infantry with, one assumes, three different MOSs and training programs would put a strain on the training base and fix a problem that doesn't really exist.[22]

In a rebuttal, however, Colonel Wass de Czege pointed out that, as a concomitant of technological advance, the conditions supporting a monolithic infantry branch no longer obtained:

> Our traditional approach has been to improve our infantry's performance on their middle ground missions.[23] Having served in mechanized, "straight leg," and airborne infantry units of the recent past like Captain Mundstock, I can say the three weren't all that different. They were designed for the middle ground primarily, with the mechanized infantry being more useful in support of tanks while the others were better in difficult terrain.
>
> With the arrival of the Bradley infantry fighting vehicle and the new light infantry organizations, however, we have a new situation. The Bradley infantry is designed to support the M1 tank, and Bradley infantry *is* significantly different from M113 infantry—or so my friends who command Bradley outfits are telling me. Although they cannot put as many soldiers on the ground for dismounted operations, they can more than make up for this with high speed mobility and heavy firepower.
>
> The new light infantry is also significantly different from the previous H-series infantry. It is truly better for mobile operations at night and in extremely difficult terrain. It cannot dig in and hold ground or assault hostile positions as well as what I call "regular" infantry. . . .
>
> Clearly, when required to do so, both Bradley and light infantry can dig in and hold a fixed position or assault an enemy strongpoint. But by organization, equipment, and training emphasis, neither is best for that role.[24]

Perhaps the most interesting feature of this debate is that it was conducted almost entirely in internalist terms. No one involved pointed out that this trichotomy of infantry missions already existed, de facto, with the other major users of infantry IFVs. The Soviets, for instance, maintained distinctive motor rifle forces equipped with BMP-series machines in tank formations versus those in motor rifle formations, which continued to use wheeled BTR-60s and their successors. The *Bundeswehr* included *Marder*, M113, and light infantry (actually mountain infantry and paratroop) units. The French, likewise, equipped infantry units variously with their AMX 10P IFV and the lighter, wheeled VAB APC, depending on their anticipated roles.

This internalist, monistic outlook was quite widespread. A prominent example of the sort of role confusion complained of by Wass de Czege—one provided by a self-described Bradley advocate—may be seen in an article by a recent former commander of the 2nd Armored Division, Major General (retired) Richard Scholtes. Scholtes considered himself a confirmed Bradley supporter:

Some individuals may be tempted to use this article as ammunition in the ongoing battle against the M-2 [*sic*] Bradley IFV. I want to make it clear that the M-2 Bradley is one of the finest vehicles and weapon systems in our inventory. If I had a choice, I would increase the number of infantry in the armored divisions by adding more M-2s and more infantrymen. With existing end strengths,[25] that may not be possible. I personally have more experience with the M-2 Bradley than the vast majority of its critics.[26]

However, he was concerned about the lack of dismountable infantry under the new Tables of Organization and Equipment:

Our 17,000-soldier armored division with ten maneuver battalions can dismount a strength of only 960 infantrymen. The paucity of dismounted riflemen, that is, infantrymen who are not platoon leaders, squad leaders, or radio/telephone operators, is even more dismal—less than 4% of the division. A World War II mechanized battalion had 234 dismounted riflemen versus the 180 of today. Our post–World War II and pre-Vietnam era division had 324 per battalion versus the 180 of today. Has the armored division's mission changed so dramatically over the years? Has the need for dismounted action decreased? I think not.[27]

In fact, it had. The contemporary U.S. armored division was optimized to defeat other armored forces, a mission McNair had always strenuously resisted in favor of pursuit and exploitation roles when overseeing the design of its World War II counterpart. In any event, a "heavy" armored division, such as Scholtes's own 2nd Armored had only three armored infantry battalions on its establishment in World War II,[28] where Scholtes himself commanded four, albeit with six tank battalions rather than three.

Scholtes went so far as to offer a radical solution, suggesting that the turrets of two BFVs per platoon be removed, for a net gain of four dismounted troops—but five "riflemen," by his reckoning. The turrets of the other two would be retained to provide a fire support element.[29]

With the current platoon organization we find infantrymen failing to dismount when the enemy situation or terrain considerations indicate it is appropriate. . . .

In defensive situations the dismounted infantrymen all too frequently are dug in close to their IFVs instead of being emplaced where their short-range weapons can have the most effect on the outcome of the battle. . . .

With the M-2/M-4 [unturreted BFV] mix, the M-2s can be positioned where they can best provide fire support to the platoon, while the M-4s can accompany the dismount squads for rapid displacement. The M-4s also provide greater carrying space for additional equipment and ammunition. An additional side benefit coming from such an M-2/M-4 mix is that it helps to get our infantry away from thinking they should fight from inside the carrier—a ridiculous idea for any infantryman.[30]

Of course, this capability had been a fundamental aim of the Army's infantry fighting vehicle development effort since 1963 or so.

However, at the same time as Bradley-equipped units were found to have trouble with conventional infantry missions, they were meeting with approbation in the armor-centered mission for which they were intended. The capabilities inherent in the M2 as a mechanized infantry vehicle, together with those of the M1 tank, renewed pressure to develop lower-level standing combined arms units than the brigade, an organizational concept that the Infantry School had dismissed as "too radical" in the course of the Divisional Restructuring Study leading to Division 86.[31]

. . . For the first time, we have an infantry fighting vehicle intentionally designed for its employment in combination with the new main battle tank. We expect the effect of the Bradley IFV and the Abrams tank, when properly employed together, to be greater than that of either system employed separately.

This effect, however, is not automatic. Events at the National Training Center have shown that units find it difficult to obtain their full potential. The speed of the M1 and M2 makes their employment harder to control. . . . The demands to get infantrymen forward to breach and clear obstacles compete with the pressing need to keep Bradley TOW launchers firing from long stand-off ranges. . . . The command and control arrangements for dismounting infantrymen, unique to the Bradley force and necessarily effected hastily in this fast-paced context, are an entirely new dimension in the employment of infantry.[32]

Even before the Bradley reached the troops there were some experiments to this effect; for instance, the 1st Brigade of the 4th Infantry Division (Mechanized) experimented with extended cross-attachment of companies between armored and mechanized infantry battalions, creating semipermanent task forces. The experiment included successful rotations through the National Training Center.[33]

Extended cross-attachment, however, for all its tactical appeal, brought serious administrative problems in an organization geared to homogeneous units. One such experiment found some seventy-three administrative functions that required special or off-line management.[34] To that end, in early 1986, the commander of II Corps obtained approval to organize the three battalions (two tank, one mechanized infantry) of the 2nd Brigade, 1st Cavalry Division at Fort Hood, Texas, under Department of the Army–approved Modified Tables of Organization and Equipment (MTOEs) as Combined Arms Maneuver Battalions (CAMBs). The two tank battalions (1-8 Cavalry and 1-32 Armor) each traded one of their four tank companies for a mechanized infantry company. The mechanized infantry battalion, 1/5 Cavalry, thus ended up as a balanced task force, with two tank, two mechanized, and one antiarmor company equipped with the M113-derived Improved TOW Vehicle (ITV). The reorganization was accomplished by September of that year, followed by a two-year test.[35]

The test was successful. The main problems were associated with the zero-sum nature of the reorganization, especially the different equipment of armored and mechanized infantry battalion Headquarters and Headquarters Companies (HHCs).[36] This situation had been a long-standing source of trouble even under the older scheme of cross-attachment, especially the mechanized infantry battalions' lesser fuel and ammunition supply facilities and the discrepancies between the two units' maintenance equipment. On a lower level, problems arose as a result of a breakdown in the parallel structures intended for armor and mechanized infantry battalions under the Division 86 structure.[37] In particular, the mechanized infantry company executive officer's lack of an IFV of his own proved to be an embarrassment. Although the executive officer was expected to be a fighter, by MTOE he was assigned to ride in an M113, with its relative lack of mobility and protection. In contrast, a tank company's executive officer commanded his own tank, enabling him to move with the other combat elements of the company and take command quickly if necessary.[38] On the other hand, anticipated training problems were not as severe as expected:

> To say that CAMB is an emotional issue in relation to branch affiliation is a gross understatement. One of the first issues invariably raised is the suspected degradation of individual training that may result from branch biases of the various CAMB leaders.
>
> A case in point is the concern over training and use of the Bradley-mounted infantryman. . . . the feeling is that, particularly under an armor-based CAMB, the 11M soldier will not receive adequate training in basic infantry fighting skills. . . . But, in reality, individual training takes place almost entirely within the purview of company command. Under CAMB, recall, nothing changes at line company level.[39]

Training problems were even arguably reduced due to the newly reemphasized focus of the mechanized infantry companies within the CAMBs, an effect predicted by Wass de Czege.[40] Instead of practicing skills such as rappelling and airmobile assaults, for instance, the CAMB mechanized infantry companies concentrated on tasks more specific to working in conjunction with tanks, such as breaching and securing major tank obstacles. The result was a refinement of the skills needed in a mechanized task force, rather than the general atrophy of dismounted skills sometimes feared by critics of the idea.[41] The general assessment of the CAMB experiment was remarkably positive. As one participant pointed out, given the time spent in training for cross-attachment prior to rotation to the National Training Center for exercises, the real question was why units should revert to pure organization the remainder of the time.[42]

The Bradley's capabilities spurred proposals for even lower level applications of the combined arms principle, suggestions being made as to the occasional feasibility of platoon and even section teams of BFVs and M1s instead of the company teams officially envisioned as the lowest combined-arms grouping. These proposals were advanced despite the problems inherent in integration due to the small-unit training incompatibilities avoided by the CAMB experiment going on at the time.[43] The major problem with this suggestion was pointed out by an armor officer, Captain Edward N. Rouse, Jr., who noted:

> . . . It is difficult enough to habitually attach companies and platoons to form task forces and company teams, let alone expand this to platoon teams.
>
> Short of developing a "combined arms" branch composed of tankers and mechanized infantrymen, the creation of platoon teams will generally bring more problems than benefits.[44]

Indeed, as predicted earlier, the complexity of the lowest-level mechanized infantry organizations was found to make proper small-unit training difficult.

> The Bradley is a lethal piece of machinery. We train the basic infantrymen how to maneuver, operate, and fight a sophisticated weapon system. We also demand that he take the EIB[45] test, maintain high levels of proficiency with the M16, M203, M60, Dragon, LAW[46] and the rest. Our soldiers practice individual movement techniques and every other "light" drill. Now add to that the Bradley Gunner's Skill Test, gunnery, company team live fires, air assaults, and any other infantry tactic you can think of, and the life of an 11M is easier to understand.
>
> Doctrine demands that the 11M fight as effectively as any other infantryman. A review of any skill manual for 11M shows not just Bradley skills but basic and advanced infantry skills as well. If "the balloon went up" tomorrow, the 11M could fight like any 11B, but could the light division soldier crawl into the turret of an M2 and kill the enemy?[47]

The first revision of *FM 7-7J*, the Bradley platoon and squad manual, was issued in February 1986. It addressed some of the above-mentioned problems of overtasking that complicated early Bradley training. To begin with, it emphasized the platoon as the basic level of mechanized infantry tactics, with the squads executing drills for conducting mounted or dismounted combat at the direction of the platoon leadership.[48]

Some effort was also made to simplify the mechanized infantry platoon structure, in particular to reduce the amount of effort needed to disperse the platoon headquarters. As discussed above, this had formerly involved the platoon sergeant moving from the first to the fourth Bradley in the platoon, changing places with a rifleman. The sergeant then displaced the squad leader of the number-four vehicle as commander. Under the new arrangement, the platoon sergeant displaced the assistant squad leader, by doctrine a qualified Bradley commander himself. This simplified the exchange, and allowed the elimination of the assistant platoon sergeant, whose main combat task was to command the headquarters' vehicle when the platoon leader was dismounted. This alteration reduced the organic strength of the platoon from thirty-three to thirty-two men. To alleviate the training problems that would otherwise be caused by the loss of an NCO, the platoon leader's gunner was promoted and designated a "master gunner," responsible for conducting the Bradley platoon's extensive mounted gunnery program. This arrangement was analogous to the administrative practice in tank platoons.[49] Also analogous to tank practice was a change in mounted movement technique. The platoon's vehicles were divided into pairs for easier control and mutual protection. This, too, was derived from armor practice. It is worth noting that the arrangement was not completely formalized, and the armor term "section" for a pair of tanks was avoided.[50]

In short, the new Bradley IFV was popular with the mechanized infantry units it equipped, although it had extensive training and logistical challenges for them. This situation was exacerbated by an accompanying doctrinal and organizational revision. Nonetheless, the mechanized infantry community successfully assimilated the new vehicle, experimenting with new techniques to exploit its capabilities in conjunction with tanks and adapting organizations, equipment, and procedures in the light of experience. Faced with an essentially similar vehicle, the armored cavalry was having similar problems; unfortunately, it faced more of a struggle.

NOTES

1. Ralph A. Hallenbeck, "Reorganize Platoon," *Infantry* 73 (November–December 1983): 12.

2. Holmes, "Light Heavyweight," 35.

3. U.S. Army Infantry School, *Special Text 7-7-1, The Mechanized Infantry Platoon and Squad* (Ft. Benning, GA: USAIS, 1980), 1.

4. The use of the term "light tank" is particularly unfortunate; it is morphologically inaccurate given the IFV's size, and has unfortunate overtones in addition (see chapter 6). As Richard Simpkin points out, "Historically, . . . light tanks have come in two kinds: good ones which were a rave in peace and a flop in war; and the rest, most recently [the Sheridan], which were a flop in peacetime too" (Richard Simpkin, "The Light Tank: A Viable Proposal?" *Military Technology* [August 1982]: 92).

5. ST 7-7-1, 1-1f.

6. Romjue, *From Active Defense to AirLand Battle*, 13–21.

7. Echelonment of forces is a key concept in Soviet-style operations, in particular offensive ones, and bears almost endless study by specialists. One such, Professor Peter Vigor of Sandhurst's Soviet Studies Center, defines the second echelon as a force held back for the purpose of achieving a specific operational goal after the first echelon has been exhausted by attrition and/or logistical deficiency. The second and subsequent echelons are distinct from reserves, which are considered units or formations a commander keeps back for contingencies, reserves being on a much smaller scale than second echelons. See P. H. Vigor, "Soviet Echelonment of Forces," in English, Addicott, and Kramers, eds. *The Mechanized Battlefield*, 15–23.

8. John L. Romjue, *A History of Army 86* (Fort Monroe, VA: Historical Office, United States Army Training and Doctrine Command, 1983) I: 66f.

9. Nicholas F. Altomare, "Bradley NET," *Infantry* 73 (July–August 1983): 29. Captain Altomare was a New Equipment Trainer during this period.

10. Robert L. Friedrich, "NET," *Infantry* 75 (September–October 1985), 33. Lieutenant Colonel Friedrich was commander of 2-41 Infantry during its reequipment with the Bradley.

11. Friedrich, "NET," 34.

12. *Ibid.*

13. Altomare, 30.

14. Ralph A. Hallenbeck, "Reorganize Platoon," *Infantry* 73 (November–December 1983): 12.

15. Tsouras, *Changing Orders*, 229.

16. The improvements in armor were of two sorts, first the adoption of advanced composite armor as in the latest Western tanks, such as that of the M1, on models subsequent to the T-72; this was dealt with by increasing the TOW warhead diameter in the Improved TOW and TOW 2 models. The other innovation was the retrofitting of older tanks with explosive reactive armor, which detonated on attack to disrupt shaped charge jets. This was dealt with first by the use of tandem charges in the TOW 2A version, arranged so that the first charge would expend the reactive armor allowing the second to penetrate the armored envelope proper, then by abandonment of axial shaped–charge attack altogether in the TOW 2B. This variant overflies the target to attack its top aspect with explosively formed penetrators.

17. Mission-Oriented Protective Posture; the individual soldiers' NBC defense equipment.

18. Bob Lessels, "Second-Generation Bradley's [*sic*] Arrive at APG for Testing," *AMC News* (November 1984): 10.

19. Huba Wass de Czege, "Three Kinds of Infantry," *Infantry* 75 (July–August 1985): 11, 13.

20. *The Mechanized Infantry Platoon and Squad, FM 7-7* (Washington, DC: U.S. Department of the Army, 1985).

21. Paul L. Conway, "Upgrade M113 Units" (letter), *Infantry* 76 (November–December 1986): 6. Major Conway, an infantry officer, was a brigade S-4 (logistics officer) in Europe during the formation's transition to the BFV.

22. Jack E. Mundstock, "Just One" (letter), *Infantry* 76 (March–April 1986): 7. Captain Mundstock was an infantry officer at Ft. Bragg, NC.

23. That is, Wass de Czege's "regular infantry" mission.

24. Huba Wass de Czege, "More on Infantry," *Infantry* 76 (September–October 1986): 13.

25. Mandated for the force structure.

26. Richard Scholtes, "Where Have All the Infantrymen Gone?" *Armed Forces Journal International* 124 (October 1986): 92.

27. *Ibid.* From the viewpoint of a commander, the situation was perhaps more alarming in view of the realities of available personnel; as one junior officer stationed in Europe noted: "even infantry battalions in Europe, where manning levels are highest, are fortunate if half of their 36 rifle squads have eight men actually working day-to-day in

each of those squads" (Terry L. Duran, "Shortcomings in New TOE," *Infantry* 76 [March–April 1986]: 39).

28. House, *Toward Combined Arms Warfare*, 109.

29. Scholtes, "Where Have All the Infantrymen Gone?" 94f.

30. *Ibid.*, 97.

31. Romjue, *Army 86*, I: 65.

32. Robert G. Bernier, "The Combined Arms Maneuver Battalion," *Armor* 97 (January–February 1988): 14f. Lieutenant Colonel Bernier was commander of 1st Battalion, 8th Cavalry, an experimental combined arms maneuver battalion (see below).

33. William A. DePalo, "Extended Cross-Attachment," *Infantry* 74 (July–August 1984): 5f. Lieutenant Colonel DePalo developed the implementation plan for the experiment and was commander of one of the battalions involved, 1/10th Infantry. See also Matthew Moten, "CAMBs: A Better Solution," *Infantry* 78 (September–October 1988): 13–15 for a concurring opinion; Captain Moten was a staff officer in the 2nd Brigade, First Cavalry and later a tank company commander in 1/5 Cavalry during the CAMB experiment.

34. Bernier, "Combined Arms Maneuver Battalion," 16.

35. *Ibid.*, 14.

36. *Ibid.*, 16.

37. Romjue, *Army 86*, I: 51.

38. Bernier, "Combined Arms Maneuver Battalion," 17.

39. *Ibid.*

40. Wass de Czege, "More on Infantry," 14.

41. Bernier, "Combined Arms Maneuver Battalion," 17.

42. *Ibid.*, 19.

43. Joseph K. Miller, "The Platoon Team," *Infantry* 76 (January–February 1986): 14–16.

44. Edward N. Rouse, Jr., "Platoon Teams" (letter), *Infantry* 76 (July–August 1986): 4.

45. Expert Infantryman's Badge.

46. Light Antitank Weapon, a disposable rocket launcher.

47. Robert S. Bobinski, "Absurd" (letter), *Infantry* 76 (November–December 1986): 5. Lieutenant Bobinski was an infantry platoon leader in the 1st Cavalry Division.

48. United States. Department of the Army. Headquarters. *The Mechanized Infantry Platoon and Squad* (Bradley), *FM 7-7J* (Washington, DC: Department of the Army, 1986), 1-1.

49. FM 7-7J, 2-4, 2-6.

50. FM 7-7J, 2-8.

Chapter 9

Armored Cavalry: Problems of Doctrine and Equipment, 1976–1989

Scouts Out! To a cavalry platoon leader in an armored cavalry regiment, that means one thing; to a scout platoon leader in a tank battalion, that means something entirely different. However, the Tables of Organization and Equipment (TO&E) under the H and J series appear to treat the two platoons as one.

—Major Barry Scribner,
operations officer,
1-64 Armor[1]

Cavalry or armored recon units do not need Winnebago M3 Bradley vehicles for scout vehicles. The people responsible for the cavalry units equipped with Bradley vehicles for scout section in the cavalry troop should be in *jail*.

—Lieutenant General (Retired) James Hollingsworth[2]

The infantry's adoption of the Bradley in its M2 guise was relatively straightforward; a generation's worth of American infantrymen had, after all, grown up with the imminent introduction of an infantry fighting vehicle. Although there were some misgivings about its effects on training and manpower, few questioned the necessity for the vehicle under the conditions in which the branch expected to conduct future operations.

The armored cavalry community was less sanguine. In the half-century-old argument between advocates of fighting cavalry and stealthy cavalry, the M3 Bradley was open to criticism from both sides. Almost from the publication of the findings of the Larkin task force, armored cavalrymen had raised dissenting voices over many of the same issues as the GAO, citing mobility problems, enhanced signatures, and inappropriate weapons fits. Despite the new emphasis on firepower in armored cavalry forces under Active Defense, the advent of a cavalry vehicle so much larger and heavier than those of previous experience was a shock to the system. Even before the first Bradley was fielded, the combination of radically new equipment and doctrine stirred controversy in the semipublic forum of the Army professional press in a way seldom seen.

Two of the first published protests, a pair of 1979 letters to the editor of the branch journal *Armor* from two NCOs are worth quoting in toto, as they distill most later objections to the M3 CFV.

Dear Sir,

I would like to know why, in this day and age when every other major NATO nation has excellent reconnaissance vehicles such as the British *Scorpion*, German *Luchs*, French *AMX-10RC*, and Canadian *Lynx* designed specifically for reconnaisance, the Army continues to saddle its Cavalry scouts with oversize, under-armed, hand-me-down *M-113*'s from the Infantry and modified Infantry Fighting Vehicles (IFV) (i.e. the Cavalry Fighting Vehicle XM-3) that are loaded down with electronic gadgetry and missile systems that the scout does not need.

I know I don't speak for all Cavalrymen, but from those I have associated with I know that a good many do agree with me. We need a *recon vehicle*, not obsolete infantry hand-me-downs or modified IFV's.

There are four areas that should be addressed in such a vehicle. They are *size*, *firepower*, *mobility*, and *simplicity of design and operation*.

Both the *M-113* and the Cavalry version of the IFV fail miserably in these categories. Both are far too large and make good targets. Both are much larger than their NATO contemporaries, except the *Luchs*. Specifically, they have much taller silhouettes, which make them easier to spot and engage, and harder to hide.

We need a vehicle that is smaller and has a lower silhouette. The *M-114* was the right idea but poorly powered and of poor reliablity [*sic*]. A vehicle of its size is what we need, and could be powered by a turbine engine. These engines have shown immense horsepower potential for their compact size. This power plant would give the vehicle the mobility to evade what it could not destroy.

The *M-113* is deficient in firepower while the IFV has more than necessary. The scout's mission is reconnaissance, *not* killing tanks. He fights only in self-defense or to extricate himself from a position. For this he does not need the TOW missile system. The most he needs is a powerful, reliable, yet simple, automatic cannon.

Lastly, he does not need half an infantry squad (the Cavalry version of the IFV is slated to have a crew of five) for a crew to accomplish its mission. Past experience in Europe and elsewhere has shown me that time to accomplish the mission seldom allows the scout the luxury of dismounted patrolling. Thus the argument for the larger crew to allow the scout to better perform this role is not valid since it will be of very little practical advantage to him. Keep the crew size down and you can keep the vehicle size down. A three-man crew is adequate.

I realize the above is wishful thinking and even if it were not we'd test it to death before adopting it anyway; but it does make for some pleasant fantasizing.
Dale T. Ewing
Staff Sergeant, Cavalry[3]

Dear Sir:

A voice from the wilderness.

I was appalled by the articles about the new Cavalry Fighting Vehicle in the May–June issue.

I remember when European cavalry units were exchanging the *M-114*s for *Sheridans* in the 3 for 5 program. At the time, critics of the program said that the *Sheridan* was too large, too expensive and too complicated to be a scout vehicle. Yet, after years of research, we have produced a "scout" vehicle that is every bit as large as a *Sheridan*, 7 tons heavier, and from all appearances every bit as expensive and complicated. I find it incredible that the Soviet Union can field an equivalent vehicle (in terms of performance and firepower), the *BMD*, 7 feet high, 15 feet long and only 9 tons in weight; and the best we can do stands better than 9 feet high, is over 20 feet long and weighs almost 24 tons! It's going to be damned hard to scout in a vehicle as large as a tank.

There is also a problem with the vehicle's armament: there's too much of it. A scout who has the capability to kill a tank at 3,000 meters is going to be tempted to shoot instead of report enemy tanks. The addition of the TOW system also places another training burden on the commander who has little enough time to train his people to scout.

The only thing I can see doing with this monster is to use it to replace *Sheridans/M-60A1*s in cavalry platoons and find a light scouting vehicle, perhaps the *XR-311*, for the scouts; but of course this won't happen.

And to introduce this vehicle in an issue where General Patton says we should keep Cavalry as light as possible....!?

Peter L. Bunce
Staff Sergeant[4]

These two letters brought a series of follow-ups. A tanker, SSG Thomas P. Currie, particularly excoriated the CFV's antitank capabilities: "Give our scouts a vehicle that can out-run and out-shoot the enemy scout, without tying him to a 'Weapons System' designed to engage main battle tanks."[5] In the same issue, an Indiana National Guard cavalryman, SGT Christopher F. Schneider, while expressing reservations about the dimensions of the CFV compared to earlier vehicles, differed on the firepower issue:

. . . A scout must be able to defeat *all* types of targets he can expect to encounter. If he can not, then other forces must be brought in to do the job and that detracts from the overall picture. Armament should be a machinegun [*sic*] for personnel, an automatic gun for light armor, and a missile for heavy armor targets. Artillery can not destroy or suppress everything and will not always be available. It is also sometimes impractical because of the short engagement times. The armament is there to protect the scout. He will not always see the enemy first. If trained properly, the scout can use this firepower to his advantage. Some of the missions a scout performs *require* the use of firepower. It is very unnerving to run up on a tank with a *M-113*.

One reason we have a problem with picking a good scout vehicle is because we have incorrectly defined the missions of a scout in relation to the present battlefield. When we do that, we will know what is needed.[6]

As if to confirm the latter comment, SSG Craig G. Mosher, writing from an infantry battalion scout platoon, differed in turn somewhat later:

> . . . [W]ill the CFV be able to fit the role of a scout vehicle? Since the CFV is as big as a house and very heavily armed, it is questionable. The CFV will carry two TOW missiles in the launcher and 10 stowed, a 25-mm cannon, a machinegun, and a squad of scouts. I see a pattern forming. Arm the scout and send him out to fight tanks and infantry. There can be no doubt that that is what will happen once the commander sees all that firepower on one vehicle, [*sic*] Now, in each scout section we will have two large, unproven vehicles that are armed to stick it out and fight with a vastly numerically superior enemy. When will we ever learn. Keep the scouts simple and equipped with a vehicle that will keep us quiet, quick, and up with the times. Scouts need the right "horses" to do our job. Don't give us a heavily-armed vehicle and then expect us to slug it out with tanks. We need to keep our cliche: "Sneakee Peakee."[7]

These letters apparently struck a nerve; they drew a considerable volume of published response. Where NCOs were the greatest critics of the M3, midlevel officers were its defenders, the most vocal of them being Major Marc A. King of the Armor Center's Directorate of Combat Developments, who contributed both a letter and an article to the March–April 1980 issue of *Armor*. Once again, his letter is worth reproducing at length for both its comprehensiveness and its mixture of realism and dogmatism.

> . . . The first point raised by SSG Bunce is vehicle size. There can be little argument that the CFV is large for a scout vehicle. The overriding consideration in relation to size was tied to the Cavalry Scout *Ad Hoc* Committee Study (1974), which concluded we needed five scouts per vehicle in order to accomplish the scout mission. In order to get five men and their associated equipment around on the battlefield, you have to provide room in the vehicle.
>
> We wanted increased survivability for the crew and that meant special armor, hence more vehicle. We needed mobility equal to or greater than the XM-1 and that required an engine and suspension configuration that took up more space.
>
> While concern for size is valid, the idea of "sneak and peek" scouting has long outlived its usefulness on the modern battlefield. Sophistication has seen to that. With heat sensors, thermal sights, people sniffers, electromagnetic signatures [*sic*], acoustical sensors, improved optics, and so forth, the fact of the matter is you cannot sneak up on anybody.
>
> Correct scouting techniques (dismounted), coupled with common sense, will be the tools the scout needs to survive—not a smaller vehicle.
>
> On the topic of vehicle armament, I can only add that part of what is currently on the vehicle was mandated by our leadership in very high places and that's what we have to live with. We face a heavy threat in Europe and "heavy" scouts make good sense. As for "shooting tanks at 3,000 meters" SSG Bunce has touched on a very sensitive point. If you think you have to shoot at every tank you see, you probably will.
>
> On the other hand, a *properly trained* scout crew (SSG Bunce's job) should possess the knowledge of when to shoot and the discipline to enforce that knowledge. The antitank capability does not equate to a dedicated antitank force. . . .

> It should be clear to tankers and cavalry men everywhere that the CFV does in fact leave something to be desired as a full-fledged scout vehicle. But the reality is that it's the only alternative to the M-113 and all we have been offered with which to do the job. Make no mistake about it—it is better than the M113 and represents the level of technology we must take to war in the 1980s if we are to win.
>
> It is time to get on board! Instead of criticizing the CFV, start thinking out how you are going to train with it and employ it in conjunction with its five-man crew. If you can't, then at least don't hinder what progress we are making. Your opinion counts—make it positive.[8]

A similar mixture, again explicitly using Bunce's letter as a jumping-off point, came just over two years later from Captain Gerald A. Halbert of the Armor School. His article, painstakingly documented, was a point-by point comparison of the Bradley in its CFV version to comparable Soviet vehicles. Curiously, the Soviet vehicles used for comparison, however, were the BMP, the BMD airborne combat vehicle, and the BTR-60 APC, of which only the first was employed in Soviet armored reconnaissance units. The article's general tone may be adequately conveyed by noting the statement made early on: "However, the CFV is smaller than one cavalry vehicle it replaces—the M60A1 tank."[9]

The debate over the suitability of the M3 as a cavalry fighting vehicle was further aggravated by the shifts in doctrine and force structure which accompanied the extensive re-equipment of the 1980s. As described above, the supersession of Active Defense as a warfighting doctrine by AirLand Battle was accompanied by new organization, "Army 86," which embraced a major overhaul of Army force structure, including a new organization for armored and mechanized divisions to succeed the ROAD concept.[10] Among other things, the force planners responsible for Division 86 had drastically reorganized the basic cavalry platoon, common to battalion scout platoons, divisional cavalry squadrons, and the cavalry troops of armored cavalry regiments. At the same time, cavalry units, especially the divisional cavalry squadron, served in many ways as "bill payers" for other enhancements within the heavy division. In accordance with the standardizing and simplifying emphases underlying Division 86, the old combined arms organization (see chapter 6) was replaced with a homogeneous one of six M3 CFVs in the ground troops both of divisional cavalry squadrons, which lost their tanks entirely, and those of armored cavalry regiments, where two M3-equipped platoons were matched with two standard four-tank platoons identical to those of other armored units. The same platoon of six CFVs served as a battalion scout platoon in both tank and mechanized infantry units. These replaced an H-series organization of ten M113s, four armed with pedestal-mounted TOW missile launchers—in accordance with the increased antitank emphasis of the Active Defense doctrine and the deficiencies of the M114—and six standard models with .50 caliber machine guns.[11]

It was this replacement of the tank by the M3 in many cavalry missions that concerned many higher level officers. These, when writing on cavalry matters, by and large implicitly accepted the Bradley as a fighting vehicle; their reservations lay in its use as the *sole* fighting vehicle in many armored cavalry subunits under the various Army 86 establishments. As with the controversy over the Sheridan during Vietnam (see chapter 3), the key datum was over what vehicle the new system replaced. In particular, there was some concern about the appropriateness of the TOW as the primary antitank weapons system under the conditions anticipated for the reconnaissance and covering force battles:

> While the M3 Cavalry Fighting Vehicle (CFV) of the AOE[12] squadron is armed with a tank-defeating TOW, the latter is a weapon best employed defensively at long range. During the conduct of reconnaissance, however, contact often occurs at short range between opposing forces, and a quick-firing, tank-defeating gun is much preferred in such circumstances. . . . Consequently, a reconnaissance mission may all too quickly become a hasty defense.[13]

Others, however, objected on the grounds that this attitude was based on an obsolete doctrinal outlook:

> In the past, "developing the situation" was synonymous with fighting for information, hence the concept of the fighting cavalry. This traditional view is based primarily on World War II combat experience. However, it has already been shown that reconnaissance was a small percentage of cavalry missions. To determine if the cavalry is capable of developing the situation without fighting, we must look beyond our combat experience to find an answer. . . . As already indicated, light elements are capable of performing reconnaissance without fighting for information.[14]

This position, however, begged the question of whether M3-equipped platoons constituted "light forces":

> Unfortunately for the "sneak and peek" advocates, the M3 CFV hardly fits this description. While it is fast and possesses excellent cross-country mobility, its powerful diesel engine and full-tracked suspension system are hardly "silent," and at 2.972 meters high it is taller than an M113 APC or an M1 tank. Nor, at 21.8 tons, can it be considered light and it can swim only after erecting a flotation barrier.[15]

This debate continued throughout the 1980s as M3s continued to enter service, with little sign of satisfactory resolution. In short, the armored cavalry community was less tractable with respect to the Bradley than was the mechanized infantry. Controversy over the relative value of combat and information-gathering in the armored cavalry mission had continued over the years. The CFV, as a large and heavily armed—but not tanklike—vehicle fell squarely between the two camps. Meanwhile, a greater public debate was taking shape over the Bradley in both its guises.

NOTES

1. Barry Scribner, "HMMWVs and Scouts: Do They Mix?" *Armor* 99 (July–August 1989): 33. Major Scribner was operations officer of the 1st Battalion, 64th Armor at the time.

2. "Remembering Some Hard-Fought Lessons of World War II (Part II)" *Armor* 98 (March–April 1989): 46.

3. Dale T. Ewing, "Why?" *Armor* 88 (March–April 1979): 2.

4. Peter L. Bunce, "Appalled," *Armor* 88 (November–December 1979): 3. Sergeant Bunce refers to George S. Patton, "Some Thoughts on Cavalry," *Armor* 88 (May–June 1979): 27–29; this article was the transcript of a speech Major General Patton delivered at a dining-in of the 2nd Armored Cavalry Regiment, Nurnberg, Germany, in December 1978. Patton emphasized the armored cavalry's economy-of-force missions in this speech, asserting that mobility was key to this quality.

5. Thomas P. Currie, "CFV Not for Scouts" (letter), *Armor* 89 (March–April 1980): 2.

6. Christopher F. Schneider, "Define the Scout's Mission," *Armor* 89 (March–April 1980): 2.

7. Craig C. Mosher, "The CFV as a Scout Vehicle Is Questioned," *Armor* 91 (January–February 1982): 3.

8. Marc A. King, "Time to Get on Board," *Armor* 89 (March–April 1980): 3.

9. Gerald A. Halbert, "The Bradly [*sic*] Versus the Opposition," *Armor* 91 (May–June 1982): 26–29.

10. The Army 86 tables of organization were later fine tuned and renamed "Army of Excellence" (AOE), but the current remarks cover both organizations.

11. Marc A. King, "Alive and Well," *Armor* 87 (September–October 1978): 36, fig.1.

12. Army of Excellence.

13. Peter S. Kindsvatter, *The Army of Excellence Divisional Cavalry Squadron—A Doctrinal Step Backward?* (Ft. Leavenworth, KS: School of Advanced Military Studies, U.S. Army Command and General Staff College, 1985), 35f. To repeat, then-Major Kindsvatter was a student at SAMS.

14. Charles L. Crow, *The J-Series Cavalry Squadron and Its Reconnaissance Mission* (Ft. Leavenworth, KS: School of Advanced Military Studies, 1985), 28f. Major Crow was a SAMS classmate of Major Kindsvatter, whose contemporary study is cited above.

15. Kindsvatter, *AOE Cav*, 55.

Chapter 10

The Army Defends the Bradley

For the record, *aluminum armor plate does NOT burn.*

—Donald R. Kennedy,
antiarmor warhead designer[1]

Controversy over the Bradley within the military was relatively muted, however, compared to public criticism of the system. As discussed previously, the Vietnam War and its fiscal aftermath had engendered massive bloc obsolescence in the materiel bases of all the services, with the Army in the worst situation due to the combination of wartime attrition in following the relative deprivation of the 1950s. When this state of affairs became a campaign issue in 1980, despite the fact that reequipment had already commenced, the prevailing post-Vietnam state of relative apathy on defense procurement matters gave way to new public (at least, journalistic and congressional) interest. Hundreds of billions of dollars and decades of policies were laid on the line. Like most public policy issues associated with the Reagan administration, the defense buildup was the subject of intense and highly polarized public debate. In this case, it was fuelled by a loose-knit reform movement that had been awaiting such an opportunity for some time.[2]

The confused, confusing, sluggish, and bureaucratized acquisition process that gave rise to the Army's stillborn panoply of advanced weapons systems in the 1960s and 1970s was by no means confined to that service. Inter- and intraservice turf wars, strategic indirection, and the multiple traumas engendered of the Vietnam era left a similar legacy for all the services. The history of the era's defense procurement effort is littered with examples of mismanagement, technological overstretch, and poorly formulated requirements. These problems delayed numerous programs and inflated their costs. Sometimes they actually resulted in their demise. The F-111 fighter-bomber and the C-5 transport airplane are only the most prominent of non-Army procurement crises of the time.

By the late 1970s, this history of failures, scandals, and close calls, together with the tendency toward introspection in public affairs following in the wake of Vietnam, left a sizeable community of interested individuals disillusioned with the status quo. These people, although outside the military mainstream of the defense procurement system, were still interested in rectifying the complex of problems with the system and related facets of the military establishment. This establishment, to their minds, showed little inclination to do so itself, due to a hierarchical culture that tended to deal with internal criticism by removing the critics whenever possible, or, more pungently:

> The practice of purging the ranks of "difficult" subordinates . . . over the years has produced a crop of senior officials long on form and short on substance. The long-term result of stifling dissent and discouraging unconventional views, while rewarding those who conform, is an officer corps that is sterile, stagnant, and predictable. Promoting clones, while purging mavericks, is tantamount to incest. We all know the possible long-term effects of generations of incest—feeblemindedness, debilitation, and insanity.[3]

This "Defense Reform" movement was more an ad hoc network than a formal organization, but it did have a degree of structure. Two men stood out as its theorists: private defense analyst Pierre Sprey and retired Air Force colonel John Boyd.

Pierre Sprey had been one of the "whiz kids" employed by Secretary of Defense Robert S. MacNamara in the 1960s, a civilian with a background in operations research and statistics. As a special assistant in the Office of the Secretary of Defense (OSD) from 1966 to 1970, Sprey had produced a number of analyses critical of technological trends within the armed forces, particularly the early generations of precision-guided munitions used in Vietnam. Although effective against some classes of targets not previously reachable, these weapons left the aircrews delivering them vulnerable as they guided the bombs to their targets.

John Boyd, on the other hand, had been a career military officer. A fighter pilot during the Korean War, Boyd had gone on to become one of the Air Force's premier tacticians, developing the theory of "energy maneuverability," which has continued to the present as the governing paradigm for U.S. air combat doctrine and equipment.[4] In the late 1960s, Boyd applied his theories to the preliminary design of the Air Force's F-X tactical fighter, later to become the F-15 Eagle. With the help of Sprey's post-MacNamara cost-effectiveness analyses, he followed this effort in the early 1970s with a similar one for the F-XX, which became the F-16. Boyd and Sprey's work on this latter project produced an aircraft radically different from its predecessors, a relatively economical "lightweight fighter" endowed with unheard-of maneuverability. This was achieved largely as a result of a bottom-up approach to the problem of fighter design and an iron-clad insistence "that technical features should be the *output* of a disciplined design trade-off process, not the *input,*"[5] a position that seriously alienated them from numerous Air Force technical constituencies. After his retirement in 1976, Boyd applied his talents to a larger arena, eventually evolving the theories of maneuver warfare which would bulk so large in the doctrinal renaissance of the U.S. military in the 1980s. Fundamental to his thinking was the conceptualization of combat actions as breaking down into a cycle consisting of four phases: observation, orientation, decision, and action. The contestant executing this "OODA loop" (also often referred to as the Boyd Cycle) the quickest had an inherent advantage in combat, because every iteration of the loop increases the difficulty in response on the part of the slower opponent until opposition becomes chaotic and confused. The French response to the German *Blitzkrieg* attacks of 1940 is the standard example of the phenomenon.

Working from this theoretical base and Sprey's operations research experience, Boyd and Sprey between them developed what amounted to an ideology of weapons system development. Its fundamental tenet was an emphasis on careful evaluation of the effective end use of an item within the framework of Boyd's theories, rather than on the effectiveness of its constituent parts. Standard practice, for instance, might be to quantify the performance of an aircraft's radar. Boyd and Sprey's paradigm would require first questioning whether the radar, and its associated expense and weight, were useful to the aircraft's mission in the first place, or even whether it might actually detract when tested against real-world conditions, whether through performance penalty, unavailability, or adverse impact on aircrew decision cycles.

Application of this viewpoint to front-line weapons systems tended to produce a consistent set of results. In the first place, they tended to emphasize simplicity, in terms of parts count, weapons fit, and mission equipment. Simplicity was not pursued for its own sake, but for its by-products. Small size allowed combat initiatives to be gained by stealth. Low running costs allowed more training to be conducted. Low acquisition costs allowed large numbers to be acquired, thus bringing the Lanchester Square Law into play.[6] To this end, the Boyd-Sprey philosophy advocated relatively limited roles for any given weapons system, so that its operators could develop expertise with their enhanced training.

In the second place, they tended to emphasize characteristics more subtle than those usually reckoned to define performance. Acceleration was valued over top speed, all-around visibility over the absolute acquisition range of sensors, and short engagement times over long engagement ranges. All these factors, which allowed the operators of weapons systems to change state faster than their opponents could, contributed to more timely completion of the OODA loop, and thus to theoretical supremacy. The archetype of Boyd's thinking was the original version of the F-16 fighter: smaller, cheaper, and slower than its predecessor, but also harder to see, more numerous, and more agile.

Within the defense community, Boyd's theories were regarded by many as almost Copernican in their scope. They can even be said to have produced their share of Galileos at secondhand; their influence upon the authors of the Army's 1982 AirLand Battle doctrine and its successive refinements has been mentioned.[7]

To carry the metaphor even farther, however, the theories at the heart of the military reform movement unfortunately also produced their share of Brunos.

This, then, was the public climate in which the Bradley entered production and service. By some measures, the Bradley, or at least the IFV concept in general, should have met with the approval of the defense reformers. Couched in the terms of the reformers themselves, the Bradley's mounted-combat facilities promised better observation and orientation on the vital small-unit level, while the closer cooperation made possible by the IFV/MBT teams the Army envisioned should have held the potential for shortening at least the "action" segment of the Boyd cycle.

On the other hand, as it emerged, the Bradley was a natural target for the defense reformers. It was a complex, expensive system, and one growing more expensive every fiscal year. It was, moreover, apparently an ad hoc, multipurpose, design, with a turbulent history, whose mission was poorly understood outside the Army. Furthermore, it was associated with the M1 main battle tank, another system that drew intensive criticism due to its cost and complexity. Finally, the Bradley was the nominal replacement for the established, reliable, inexpensive, simple, and protean M113 armored personnel carrier, used by dozens of countries and built in the tens of thousands.

The word "nominal" is important here. The M2 was intended to be a one-for-one replacement for the M113 within the line companies of mechanized infantry battalions; however, it was not a replacement for the older machine and its variants in all its support roles. Nor, as has been shown, was the infantry fighting vehicle's doctrinal role analogous to that of the armored personnel carrier. Unfortunately, the Army's 7- and 71-series doctrinal publications tended to muddy this distinction by prescribing similar roles to both types in combat (see chapter 8).

This situation was ideal for the defense reform movement. In the face of almost constant frustration by an elephantine defense bureaucracy, the reformers made a specialty of press relations. This was both a strength and a weakness; many in the reform movement, whether executive, legislative, or private-sector workers, were wont to play to the shortcomings of mainstream journalism as thoroughly as they flailed at those of the military-industrial complex. They tended to simplify problems and take cheap shots at complex problems to the point that the reasoned arguments behind the reform movement often began to appear to be a sort of nostalgia cult built around the *Wehrmacht*, the P-51, and the draft.[8] One former Army public affairs officer, sympathetic to the problems of general-interest journalism describes the problem thus:

> That the army functioned perfectly normally this week, that it is in general adequately trained, sufficiently equipped, and mostly in good spirits is dull, uninteresting, and unprintable. It does not . . . sell newspapers. . . . What the press is looking for . . . [is] conflict, crime, fraud, weapons that do not perform as advertised, waste and abuse.[9]

Or, in the words of one less sympathetic defense journalist, syndicated columnist Fred Reed:

> The techniques of the reformers, and of others like them, are simple. They acquire, say, an internal Pentagon document that mentions or seems to imply problems in the development of a weapon. Difficulties occur in the development of anything complicated, civilian or military, but the reporters do not seem to know this. . . .
>
> These are tempting stories. . . . They are wonderful copy, and they give the inexperienced reporter a delightful feeling of having penetrated a mystery.
>
> In contrast, the Pentagon—stodgy, terrified of saying the wrong thing, bureaucratized to the gills, and not journalistically sophisticated—takes days or weeks to answer and does a poor job. The reformers get the ink; it is not unusual to see half a dozen reformers, and no one else, credited as sources. The reporter has no idea that he or she has been used like a wrench.[10]

The Bradley (as it was to become) was also on the spot at the wrong time. Although the network of defense reformers had coalesced by the mid-1970s, defense reform did not become become a journalistic staple until the end of the decade, with the publication of James Fallows's "The Muscle-Bound Superpower" in the October 1979 issue of the *Atlantic*, expanded to the book *National Defense* in 1981. These were followed by a wide variety of pieces on almost every aspect of defense, and for the reasons discussed above, a fair number of them focused on the Bradley. Many of these stories were focused on policy and politics. These were often scathing, but were also often excellent reviews of the program to date. It is certainly fair to say that the administrative and political history of the program and its precursors carried a great deal of baggage to be accounted for.[11] In matters of technology and operations, however, there were problems, as reporters tended to display their lack of military and technical experience to poor effect. In Fred Reed's words: "If journalists had any notion how stupid they seem as they stand around a tank peering in truculent subconscious puzzlement, asking stupid questions that make everyone else cringe, they would wear ski masks."[12]

An early and widely-distributed representative of this sort, William Boly's "The $13 Billion Dollar Dud," appeared in the February, 1983 issue of *California* magazine, a regional organ; it was later picked up by the *Reader's Digest,*[13] greatly increasing its exposure. Boly's story, researched during the early phases of Bradley production in 1982, was a straightforward attack on the system; the article is being treated at some length here as it is in many ways the archetype of general-periodical technical critiques of the Bradley. Few later treatments are as comprehensive, but almost all touch on the same issues.[14] Among other things, Boly managed to criticize the Bradley for being too large and too small within the space of a page;[15] to make capital of the fact that systems still in development fail in tests much of the time;[16] and to declare that it was almost impossible for the Bradley gunner to acquire targets on the basis of a test ride, without noting that this task is, in fact, the job of the commander.[17] Despite this, Boly uncritically reports defense consultant and reformer Paul Hoven's endorsement of the FMC AIFV (see chapter 5) as an inexpensive alternative to the M2. That vehicle, however, actually does have a one-man weapons station that leaves the gunner responsible for acquiring many targets.

"The $13 Billion Dud" also broached the "vaporifics" issue, which addressed the Bradley's protection level. Hoven, for example, asked if it was wise to "take a light tank, which is what that thing really is, fill it with soldiers and put it on the offensive the way the army is talking, and all I see is a bunch of dead people."[18] Such comments were simplistic in that they ignored the Army's actual doctrine for the IFV's employment. This called for the Bradleys to stay two hundred to four hundred meters behind tanks in the armored advance whenever possible, noting that "the [mechanized infantry] platoon will seldom lead tanks mounted." Like any armored vehicle, Bradleys were expected to exploit terrain for concealment and protection as well, thus the criticism in many quarters of their height.[19]

The feasibility of the doctrine itself, on the other hand, was certainly open to question. For instance, as mentioned in chapter 7, Brigadier Richard Simpkin, hardly an unsympathetic observer, points out that in close country—certainly in the forests and urban areas of Europe—IFVs may often have to lead. Tanks have become increasingly unwieldy and their armament more specialized for fighting other tanks. This brings IFVs into greater and greater prominence, with their dismountable troops and handier weapons; this, in turn, places a premium on their protection from direct fire.[20] This protection was limited in the case of the Bradley, largely because of the insistance on perpetuating APC characteristics, in particular airportability and amphibious capability, in the newer vehicle. Unfortunately, the relentless pursuit of soundbites largely bypassed such subtleties.

Boly's article also brought up another aspect, that the Bradley's aluminum armor might actually be a danger to the personnel in the vehicle. This charge arose from an interview with an anonymous Washington defense consultant, who in turn quoted a 1980 report from the British Chemical Defense Establishment, Porton Downs.[21] The report was alleged to claim that when finely divided under the action of the jets of shaped charge antitank warheads—that is, those of the bulk of infantry antitank weapons—the aluminum armor of the Bradley would react energetically with ambient oxygen, producing heat, brilliant light, and additional overpressures to the obvious detriment of the crew, along with dangerously high levels of toxic oxides of nitrogen. In support of this claim, the anonymous informant pointed out that the crews of U.S. and Israeli M113s, made of the same 5083 aluminum alloy as the main hull of the Bradley, habitually rode on top of the vehicles in Vietnam, in the Yom Kippur War, and in Lebanon; and that the Israelis found that troops in M113s were four times as likely to become casualties as those riding in half-tracks, which the IDF used in considerable numbers.[22] These data were not entirely relevant. In the case of Vietnam, the primary threat was from mines, ambush, and an adverse climate (as discussed in chapter 3) rather than artillery. This made riding on top of the M113 a sensible choice; on a battlefield where the enemy proposed to mass eighty to one hundred guns per kilometer of front as in Soviet doctrine,[23] the choice would be different.

Similarly, the comparison of the M113 and the half-track is prejudiced by the differing uses to which APCs and half-tracks were put. Israeli M113s, equipped much like Vietnam-era ACAVs, were used, IFV-like, in direct support of tanks. They were, in fact, used more aggressively than even Bradley doctrine envisioned. Half-tracks, on the other hand, were more likely to be in secondary roles.[24] Comparisons with experience with the M113 also ignored the fact that the M2 was designed to be more resistant to attack by mines than the M113, through a combination of higher ground clearance, extra belly armor, and void space.

The reaction to this charge by the Bradley's defenders was interesting. In most cases, they distorted the charge itself. Ignoring the complexities of the vaporific effect, they declared that aluminum armor does not burn.[25] Undersecretary of the Army James R. Ambrose went so far as to offer a $10,000 reward to anyone who could set aluminum plate on fire.[26] Ambrose's money was safe, to be sure, but that was not the allegation being made.

This evasion was unfortunate, as the issue could be discussed directly to the defenders' credit. One qualified commentator, shaped-charge designer and terminal ballistician, Donald R. Kennedy did so, albeit accompanied by the dismissal that heads this chapter. In a July, 1983 article for *Armor*, Kennedy described the defeat of armor by shaped charge attack in terms just as harrowing as, and much more precise than, any used by the defense reformers:

> —The internal explosion overpressures may range from as little as a third to as much as three or more atmospheres (e.g., 5–50 pounds per square inch[)] with durations up to a [*sic*] 100 milliseconds, which can result in trauma ranging from ear drum damage to major injury. The lungs and other body cavities are particularly affected by excess overpressure. The suddenly-developed pressure may also throw the occupants about the interior of the vehicle's crew compartment with possible injuries resulting from collisions with hard objects within the vehicle.[27]

He then proceeded to discuss the actual details of the vaporific effect in as much detail as security regulations and a professional but largely nontechnical readership would allow. In the process, Kennedy raised a variety of points glossed over by Boly's unnamed source. Perhaps most important, he quoted a personal communication with Porton scientist F. P. Watkins, to the effect that one of the most troubling vaporific effects, the production of high concentrations of oxides of nitrogen, was particularly associated with large-caliber warheads, such as those of medium and heavy antitank guided weapons, rather than the more common rocket-propelled grenades raised as threats in reformer-driven stories in the general press. Similarly, Kennedy pointed out that vaporific effects manifested themselves most strongly under extreme overmatching conditions; that is, those capable of disrupting vehicle and occupants, aluminum or not. Numerous RPG attacks on M113s in Vietnam had never resulted in any reports of aluminum vapor effects. He also pointed out that steel armor could also produce vaporific effects under extreme overmatching conditions. Although it was less prone to this behavior, it produced more dangerous spall (inward-directed high-velocity fragments) when penetrated.[28] Kennedy concluded by discussing what he regarded as the real killers in shaped charge penetrations, secondary fires and explosions from fuel and ammunition, and the methods being developed to palliate them. These included "spall liners" of various substances—Du Pont's Kevlar™ polyaramid fiber being the newest and most promising—inside the main armored envelope to stop impinging fragments, better fuel and ammunition arrangements of the sort employed on the M1 but not the Bradley, and automatic fire-suppression systems used in both types.[29]

However diverting this public controversy was, it is important to remember that not all the defense reform efforts were conducted in the public prints; many of the reformers were serving officers and bureaucrats. One such was Air Force Colonel James Burton, one of Boyd and Sprey's most avid disciples and collaborators.[30] Colonel Burton was an Air Force Academy graduate and KC-135 tanker pilot by original training and later a graduate of the Industrial College of the Armed Forces; he had spent much of his career in the Pentagon in a variety of procurement-related capacities, culminating in assignments as military assistant to three consecutive assistant secretaries of the Air Force in charge of procurement. In 1982, Burton moved to OSD, on the staff of the director of defense test and evaluation. Burton was dissatisfied with the state of weapons testing, which he considered stylized: based more on meeting published requirements than the functionality of the requirements themselves; atomistically focused on components, rather than whole systems; and overly reliant upon computer simulations that he felt were of dubious fidelity. Burton pushed for realism in testing, and a focus on casualty prevention in weapons design rather than system survivability.[31] Burton's ideal was realistic, "all-up" live fire testing, using actual Soviet munitions against combat-loaded examples of major U.S. weapons systems, such as the M1 tank or the F-16 fighter. Also, most assuredly, the Bradley.

Burton claimed not to be a critic of the vehicle,[32] but given his philosophical orientation toward casualty prevention, the Bradley was a natural target of his scrutiny. In his words, "in the event of war, between 50,000 and 70,000 infantrymen would ride into combat in Bradleys."[33] Although the Navy could not be convinced to participate in Burton's effort, the Army and Air Force agreed, and the Joint Live Fire Testing Program was underway by spring 1984. Friction almost immediately developed between Burton and the Army agency delegated to perform the tests, Aberdeen Proving Ground's Ballistic Research Laboratory (BRL). BRL, which had conducted prior vulnerability tests on the Bradley while the vehicle was in development, was particularly interested in improving the computer models Burton so deplored. The laboratory hoped to build a superior database from the live fire tests, and thus preferred controlled, "building-block" experiments measuring weapons effects on components. Burton's vision therefore clashed with BRL's, as his approach could not produce statistically significant results without expending terrific numbers of vehicles.

To Burton, this was in fact the point. BRL, however, felt that its previous testing and the established requirements for the vehicle made many of the tests Burton advocated, such as firing heavy antitank missiles and tank main gun rounds at the vehicle, foregone conclusions.[34] BRL was particularly interested in marginal but proliferated threats, such as shoulder-fired antitank rockets, and their potential for inducing secondary fires and explosions in the ammo-packed Bradley troop and fighting compartments. BRL and the Army thus talked themselves into a position superficially reminiscent to the Navy Bureau of Ordnance's position on torpedoes in the early part of World War II, when the Bureau refused to test its weapons on the grounds of expense and previous piecemeal experiments. There was, though, a crucial difference: BRL was adamant that the tests were superfluous in that they considered it obvious that the Bradley would *fail* under test, given that the protocols Burton proposed included obvious overmatches well beyond the requirements to which the vehicle had been built. Burton chose to regard this difference in approach as dishonest. His relations with BRL thus quickly soured, with Laboratory officials concluding in turn that Burton was "'driven by a need to show a videotape of a catastrophic fire' aboard a Bradley."[35]

Nonetheless, after tortured negotiations over the protocols to be followed, the tests commenced in early 1984. They were to be conducted in two phases. Phase I was meant to establish the vulnerabilities of current vehicles. It was completed by October 1985, although fewer shots were fired than planned, as some overmatches—attacks that would obviously destroy the vehicle—were conceded,[36] and a missed shot completely destroyed the second test vehicle due to secondary explosions. The tests, predictably, showed that the 25-ton vehicle was highly vulnerable to weapons designed to destroy tanks of twice that weight. It also showed that the extensive internal ammunition stores were, in fact, highly vulnerable, although the effect could not be reproduced reliably; indeed, some ammunition fires unexpectedly extinguished themselves.[37]

The testing was marked by continuing disagreement over details of the testing protocols, as BRL continued to collect datapoints for modeling while Burton pushed for holistic, if astatistical realism (aside from the BFV being a static, passive, and exposed target) and considered any departure from his program to be prima facie evidence of fraud.[38] This conflict, for instance, led to Burton's lodging protests over BRL technicians' wetting of the uniforms on dummy crewmen placed in the test vehicles for one test (the uniforms gave off vapors that BRL contended would throw off tests of the Bradley's vaporific behavior) and of the Laboratory's specifying a relatively uncommon Rumanian RPG warhead (caliber undisclosed, but apparently 73mm, the caliber also used in the BMP-1 main gun and a common Soviet recoilless rifle[39]) rather than conducting the test with the more commonly encountered 85mm projectile.[40]

The different parties involved interpreted the results to suit themselves. Colonel Burton played up the potential for casualties; BRL emphasized systems vulnerability. For instance, Burton claimed that BRL had insisted on testing the M3 CFV in preference to the M2 IFV because the test would show fewer simulated casualties. BRL claimed that the M3, with almost twice the load of missiles and 25mm ammunition, would be a harsher test because of the increased chance of secondary explosions.[41] The Army Chief of Staff's office, in response, issued a white paper which reasonably pointed out that the Bradley was not intended to sit out in the field and be shot at; rather, it would be maneuvering in concert with other forces, exploiting terrain features for protection, and, of course, firing back. The white paper concluded that the tests would yield valuable data for future survivability improvements, and somewhat disingenuously asserted that, in any event, the Bradley was far superior to the M113 and BMP.[42] Congress picked up interest in the program as well, with hearings on the testing program held in January 1986, as well as further oversight hearings for mandatory progress reports in February and March.

Discussions for Phase II, which was intended to test improvements to the Bradley based on Phase I results, commenced during February 1986. Once again, Burton and the BRL officials were fundamentally at odds on the approach to be taken. Burton wanted a Bradley modified to separate the infantry squad from the fuel and ammunition, but carrying little in the way of additional armor save minor protection from small arms fire and artillery fragments. BRL, on the other hand, wanted to test a vehicle with additional armor, including explosive reactive armor.

Once again, the acrimony was increased by misunderstanding. BRL asserted it had already studied external ammunition storage and found it wanting due to the increased chance of destruction of the ammunition by small-arms and artillery fire[43] and lack of protected access by the crew. Burton claimed that his configuration merely represented a proof of concept; he envisioned production vehicles being fitted with externally vented, internally accessible compartments like those of the M1 tank, a feature whose lack he professed to find baffling.[44] Eventually, Burton agreed to competitive tests of the BRL High Survivability Testbed and his own Minimum Casualty Baseline Vehicle, which the Army preferred to call the Advanced Survivability Testbed.

Testing protocols again became an issue. Burton insisted on shots at random points on the vehicle from random ranges, at random angles of attack, to exclude possible biases. BRL wanted preplanned shots to gather the maximum information about selected vulnerabilites while preventing expensive duplication. The issue was still unresolved when Phase II testing of the BRL vehicle commenced late in March 1986, with predetermined shot lines. This was unsatisfactory to Burton, and became more so when he discovered that the BRL testers had moved two 5-gallon water cans, part of the Bradley's standard equipment, from the bottom rear to a location near the vehicle's center of mass, where eight of the thirty-four planned shots would pass.[45] BRL's officials claimed the change had been recommended two years earlier,[46] but for Burton, it was the last straw, especially combined with Army delays in fabrication and testing of his Minimum Casualty Baseline Vehicle. Burton launched a memo listing what he saw as irregularities in the testing program. As a result of these disagreements, the Phase II tests were suspended.

At about the same time, Burton received orders transferring him out of Washington, and away from testing. Rather than accept this evidently punitive reaction, he retired after twenty-seven years service. This combination of events triggered an investigation by the House Armed Services Committee, whose staff found the tests to be ill-defined, though useful. The committee further found that the irregularities cited by Colonel Burton were due to "a long-standing fundamental disagreement over testing methodology and, more importantly, the inability of OSD and the Army to reach an agreement on how the test is to be conducted. . . . The Army has complied with many of Colonel Burton's issues of concern over the past several years."[47] At the same time, the committee staff report recommended that the committee direct that the Department of Defense Inspector-General investigate Burton's proposed transfer.[48]

> During the period 15 June through 15 July 1987, the [U.S. Army Infantry] Board conducted a concept evaluation program (CEP) test of the effects of the BFVS Advanced Survivability Test Bed [*sic*] (ASTB) modification on the operational effectiveness of the BFVS. The ASTB modifications included increased armor protection applied to the hull and turret; a spall liner applied to the interior of the vehicle; restowage of ammunition . . . external stowage compartments for the 25mm ammunition and TOW missiles; external fuel cells; the relocation/redesign of fire suppression handles; and a reduction in the number of firing port weapons from six to five.[49]

This announcement was part of the resumption of Phase II. Burton's actions had not all been in vain; the resumed tests were conducted by the Army Test and Evaluation Command, and involved random shots at all-up vehicles under close Congressional scrutiny.[50]

In short, opposition to the Bradley also continued outside the Army in the context of a broader military reform movement. This group seized upon perceived shortcomings in the Bradley's protection to attack the program and thereby the defense procurement and doctrine establishments. The resulting controversy—and Colonel Burton's display of moral courage—accelerated the development of enhanced protective systems for the Bradley. In 1988, a new model of the Bradley entered production with numerous survivability enhancements added. They were, however, more BRL's than Burton's.

NOTES

1. Donald R. Kennedy, "Improving Combat Crew Survivability," *Armor* 92 (July–August 1983): 18.

2. The literature of the defense reform movement of the 1980s is notably broad and inchoate, with no scholarly synthesis readily available. The following account derives substantially from James G. Burton, *The Pentagon Wars: Reformers Challenge the Old Guard* (Annapolis: Naval Institute Press, 1993), chapters 1 through 7. Although obviously partisan, Burton's account appears to be the best available summary.

3. James G. Burton, *Pentagon Wars*, 15.

4. Energy maneuverability theory uses the relationship between the altitude and kinetic energy of a given aircraft as a basis for tactics in air combat maneuvering (*ibid.*, 10).

5. *Ibid.*, 15.

6. This statistical law states that in an m-on-n engagement, the fighting value of a discrete unit increases linearly, while that of a group increases by the square of the number of discrete units, so that, for instance, in a two-on-one engagement, the singleton must be four times as powerful to prevail.

7. They brought forth a similar revolution to the Marines under their commandant, General Alfred Gray, as well as exposing a wide variety of military officers, Congressmen and their staffers, and procurement bureaucrats to new thinking, which many of them attempted to apply. The Copernican metaphor seems particularly apt here, since the institutions involved were nothing if not Ptolomaic, presenting as they did a picture of layer on layer of endless, iterated circular motion.

8. Gary Hart and William S. Lind, *America Can Win* (Washington, DC: Adler and Adler, 1986) is perhaps the most egregious example of this tendency, although even less polemical works drawing on the military reform movement, such as James Fallows, *National Defense* (New York: Random House, 1981), exhibit it as well. The term "*Wehmacht* penis envy," coined by Professor John J. Mearsheimer of the University of Chicago political science department (see Daniel P. Bolger, "Maneuver Warfare Reconsidered," 39 n. 23 in *Maneuver Warfare: An Anthology,* edited by Richard D. Hooker, Jr. [Novato, CA: Presidio, 1993]), while evocative, is arguably simplistic.

9. Philip E. Soucy, "Media Coverage of the Military: A Soldier's Critique," 114, in *Defense Beat: The Dilemmas of Defense Coverage,* edited by Loren B. Thompson (New York: Lexington, 1991). Soucy had been a spokesman for the Department of the Army.

10. Fred Reed, "A Culture of Incompetence: Why the Daily Press Covers Defense So Poorly," 131, in *Defense Beat: The Dilemmas of Defense Coverage,* edited by Loren B. Thompson (New York: Lexington, 1991).

11. A useful example is John J. Fialka, "Embattled Weapon: A New Troop Carrier Is Remarkable—So Are Its History and Cost," *Wall Street Journal*, 17 February 1982.

12. Reed, "Culture of Incompetence," 188f., n. 5.

13. William Boly, "The Army's $11-Billion Deathtrap." *Reader's Digest* 123 (August 1983): 125–28.

14. Cf. Gary Hart and William Lind, *America Can Win*, 21–22, 43–46; also James Coates and Michael Kilian, *Heavy Losses: The Dangerous Decline of American Defense* (New York: Viking, 1985), 157f.; and "The Winds of Reform," *Time*, 7 March 1983, 12–30 for some examples. Much of this sameness may be explained by the fact that Dina Rasor's Project on Military Procurement was acting as a clearinghouse between the working elements of the defense reform movement and the press. An inside account of the project's history may be found in Rasor's memoir, *The Pentagon Underground* (New York: Times Books, 1985). For the canonical defense of the Bradley at this time, see former Bradley program manager Brigadier General (Retired) Philip L. Bolté, "Where Were They When the Paper Was Blank? A Response to the Critics of the Bradley Fighting Vehicle," *National Defense* (April 1982): 11–12, 52. One matter-of-fact answer to General Bolté's rhetorical question is "in elementary school"—a vital factor in understanding a number of the Bradley's technical and political problems. That is an indictment of the Army, rather than of the press.

15. William Boly, "The $13 Billion Dud," *California* (February 1983): 65.

16. *Ibid.*, 66; especially piquant given that successful tests would later give rise to charges of fraud. See below.

17. *Ibid.*, 66f; cf. also Fred Reed's comment: "An inexperienced reporter looks briefly at a weapon whose manual he or she has never read, whose principles he or she barely grasps, and decides that if he or she cannot understand it instantly, neither can a soldier who has trained with it for six months" (Reed, "Culture of Incompetence," 127f.).

18. Boly, "Dud," 67.

19. ST 7-7-1, 3-21; the Army muddled this issue somewhat by providing comic-book-style, out-of-scale graphics for the relevant doctrinal publications that appear to show groups of tanks and infantry vehicles in much tighter groupings, even in publications intended for M113-mounted units; see FM 7-7 (1985), figure 4-7 for one such example. Additional muddle came from the pivotal 1979 Senate Armed Services Committee hearing where a MICV program manager stated, in a poorly conceived rhetorical flourish, that ". . . the Army feels rather strongly that there is an urgent requirement for infantry and cavalry fighting vehicles to fight side by side with the M1 tank"; the statement echoed through the years (Tom Donnelly, "How the Bradley Became a Target," *Military Forum* [March 1988]: 51).

20. Simpkin, *Mechanized Infantry*, 66.

21. Boly, "Dud," 127. This document has proven difficult to trace. Neither reformers nor Bradley partisans at the time appear to have given useful citations or even quoted it directly. Burton (p.281 n.17) does cite vaporific effects information from: "F. P. Watkins, Canadian Defense Establishment, 'The UK Wound Ballistics Research Programe [*sic*] on Behind Armour Effects,' report presented at The Technical Cooperation Program (TTCP), meeting 8 of Sub-Group W, Technical Panel W-1, Terminal Effects, 15–24 October 1980, Valcartier, Quebec, Canada."

22. Boly, "Dud," 127.

23. Isby, *Weapons and Tactics,* 162.

24. Edwin L. Kennedy, Jr., "Israeli M113s," *Infantry* 74 (July–August 1984): 7. Captain Kennedy, an Infantry officer, had also attended the Israeli Armor Commander's Course.

25. Donald R. Kennedy, "Improving Combat Crew Survivability," *Armor* 92 (July–August 1983): 18; Soucy, "Military Coverage," 120; Reed, "Culture of Incompetence," 130 (both collected in Thompson, ed., *The Defense Beat*), among many others.

26. Dennis Steele, "Taking on the Hits," *Soldiers* 41 (August 1986): 11.

27. Kennedy, "Improving Combat Crew Survivability," 17. Note that this is one item from a list that goes on for over a column.

28. *Ibid.*, 17–19. Burton (n. 21 above) places Watkins at the Canadian Defense Establishment.

29. *Ibid.*, 20f.

30. The backbone of the following account is provided by William Rosenau, *Live Fire: Testing the U.S. Army's Bradley Fighting Vehicle* (Cambridge, MA: John F. Kennedy School of Government, 1988), a Kennedy school instructional case funded by the Department of the Army, and Burton's memoir, *The Pentagon Wars*, cited above.

31. At first glance this is a fairly subtle distinction; it can best be described as an emphasis on preserving the operators of a weapons system—which are, after all, the hardest component to replace—rather than on keeping the components themselves functioning in the face of ballistic insult. The canonical example of this philosophy may be found in the design of the Israeli *Merkava* main battle tank, where, for instance, the power pack is at the front of the vehicle—unlike any other main battle tank—to serve as part of the protective system for the crew, as do elements of the suspension and other components. See Ogorkiewicz, *Technology of Tanks*, I: 59f.

32. Rosenau, *Live Fire Testing,* 9f., n. 23; note, however, that in his memoir Burton describes the Bradley in stereotypical defense-reform terms (Burton, 132–34). An egregious example is his characterization of the M113 as an eleven-man vehicle (that is, including the two-man crew in his reckoning), then stating that the Bradley dismounts six (that is, excluding the BFV's three-man crew). Such a statement also neglects the reduction of M113-equipped squads to nine men in 1983, and the fact that pre-Bradley mechanized infantry doctrine built a "vehicle team," including the squad's heavier weapons, around the APC crew, generally leaving a "dismount team" of six even at full strength (*FM 7-7, The Mechanized Infantry Platoon and Squad* [Washington, DC: Headquarters, Department of the Army, 1977], 2-3). As usual, where the reformers saw a failure to achieve an ideal, the Army saw an improvement over the existing situation.

33. Burton, *Pentagon Wars*, 132; this is, like many defense reformer statistics, bloated for dramatic effect. Were the entire projected Bradley production run of 6,882 vehicles committed to battle at once (with no training machines, war reserves, or maintainance and logistical float) and were they all nine-man infantry fighting vehicles, fully manned (approximately 40 percent were five-man cavalry fighting vehicles) the sum would be 61,938 men. More realistically, five heavy division equivalents and two armored cavalry regiments in Europe would deploy approximately 1,800 Bradleys in the unlikely event of full strength, for a figure of 13,000–14,000 men, applying the 40 percent figure for CFVs and generously counting the armored cavalry regiments as equivalent to a sixth division; still quite a large number, but not what was claimed.

34. U.S. Congress. House. Committee on Armed Services. *HASC Staff Inquiry into the Bradley Live Fire Test Program* (Washington, DC: HASC, 1986), 7a.

35. Rosenau, *Live Fire Testing*, 9. Quote attributed to Gary Holloway, BRL engineer and program manager for the Bradley live-fire tests.

36. U.S. Congress. General Accounting Office. *Bradley Vehicle: Concerns about the Army's Vulnerability Testing* (Washington, DC: GAO, 1986), 6.

37. Burton, *Pentagon Wars*, 162.

38. Rosenau, *Live Fire Testing,* 8f.

39. Isby, *Weapons and Tactics,* s.vv. "RPG-7," "SPG-9."

40. Rosenau, *Live Fire Testing,* 8; Burton, *Pentagon Wars*, 138.

41. Rosenau, *Live Fire Testing,* 11 n. 27. Cf. Burton, *Pentagon Wars*, 169; Colonel Burton claims that BRL's selection of the M3 was chicanery rather than a difference in emphasis. Whenever Burton cites a figure in his memoirs for how much ammunition a Bradley carries (e.g., 161, 166) he uses the larger M3 figure.

42. U.S. Army. Chief of Staff. *Bradley Fighting Vehicle: White Paper 1986* (Washington, DC: U.S. Department of the Army, 1986).

43. Rosenau, *Live Fire Testing,* 12.

44. Burton, *Pentagon Wars*, 190.

45. *HASC Staff*, i.

46. Rosenau, *Live Fire Testing,* 13.

47. *HASC Staff*, ix.

48. *HASC Staff*, xi.

49. "Infantry News," *Infantry* 77 (November–December 1987): 7.

50. Burton, *Pentagon Wars*, 234.

Chapter 11

Adaptation and Application: The Bradley in Service, 1988–

You have to understand, while we fight as a combined arms team, we plan and budget by line item.

—Major General Peter M. McVey,
Program Executive Officer,
Armored Systems Modernization,
TACOM[1]

The A2 series of Bradley Fighting Vehicles that emerged from the FMC production line in May 1988 was significantly different from the earlier versions. The A1 vehicles had featured minor upgrades to the offensive systems; the A2s featured radical ones to the defensive system. The basic change was a straightforward thickening of the armor. Steel appliqué plates on the turret and hull front raised the protection level from the A1 standard of resistance to the 14.5mm machine gun found on the Soviet BTR-series APCs to resistance to the 30mm cannon found on the BMP-2. Larger, solid plates replaced the spaced laminates protecting the sides of the earlier vehicles as well, providing a similar standard of protection. This heavier structure, in turn, allowed the addition of reactive armor against shaped charge attacks from rocket-propelled grenades and antitank missiles—the first use of such protection in a U.S. vehicle. All this augmentation added about seven tons to the combat weight of the BFV, making a total of almost 32 tons. To retain mobility, the power train was upgraded early in the production; its output rose to 600hp, up from 500. The torsion bars were beefed up as well, to compensate for the additional weight.[2] Similarly, the swim curtain was extended to cope with the need to increase displacement.[3] Swimming requirements, in addition, were relaxed in view of the need for enhanced protection, and the recognition, some thirty years after the *Bundeswehr* had reached the same conclusion, that infantry fighting vehicles, being tied tactically to the tank, had limited chances to exploit amphibious capabilities. Major General Peter M. McVey, who was in charge of the Army Tank-Automotive Command's close combat vehicle projects at the time, when asked if cost considerations had limited the Bradley's level of protection, replied:

> No, I think it's more doctrinal than anything else. We had a requirement to swim the Bradleys. We met the requirements for armor protection. The trouble is the world doesn't stand still. As it got more lethal out there, we were going to put applique armor on. Then you get the swim problem. The Army backed off the ability to swim. Therefore we increased the armor protection on the Bradley.
>
> Like everything else, it was a trade off. As long as they wanted to swim, and you're putting excess weight on there, we could swim as long as we could grow those extension curtains. If you have a tear in one of those things you'd swim like a rock. So, the Infantry Center, supported by TRADOC and the DCSOPS,[4] said: "We'll back off the swim capability." The Bradley goes where the Abrams goes; the Abrams doesn't swim. It was good rationale. It wasn't convoluted. When we did that, we increased the capability of the Bradley.[5]

These gross improvements were accompanied by subtler ones. Ammunition and equipment were restowed in accordance with live-fire test data, and the dismount team member formerly seated behind the driver in the cramped space beside the turret was moved to a new position behind the turret. Additional internal protection was provided by Kevlar spall liners. These liners, designed to catch the secondary fragments produced by penetrating shaped-charge jets and, to a lesser extent, other projectiles, were an improvement promoted by Burton and already in use on the upgraded M113A3s entering service.[6] The APC's straightforward, boxy internal geometry made this a relatively simple addition, whereas the convoluted interior of the BFV did not. One other survivability enhancement favored by Burton and pioneered on the M113, the removal of the BFV's fuel supply to external armored cells, was not adopted; nor was the extensive M1-style externally-vented ammunition stowage he advocated adopted, although some munitions were moved outside the rear compartment.

These changes had an important secondary effect: the greater coverage of the new side armor and the structural demands of reactive armor meant that the side firing ports, and their accompanying periscopes and firing port weapons, were omitted. As a result, the capacity of the dismount element of the BFV squad to fight or observe while mounted, originally a fundamental feature of the infantry fighting vehicle idea, was almost eliminated. The firing ports, M231 Firing Port Weapons, and vision blocks to the rear remained, however.

While the A2 vehicles were developed, production and fielding of the A1 series proceeded, and the Army continued to accommodate the Bradley within the Army 86/AOE (Army of Excellence) organizational framework. A new edition of *FM 7-7J* was prepared to reflect experience gained from the first few years' experience with the BFV. The manual was issued in 1993; however, its most important change was announced in January 1990. The mechanized infantry platoon was reorganized to form separate organizations for vehicle crews and dismount teams, a sharp break from mechanized infantry practice since the ROAD reorganization of the early 1960s, and a sharp deviation from the intent of training programs since the Bradley was first issued to mechanized infantry units. The merits of somehow separating the two functions had been discussed for some time, however, owing to the demands of training squads and platoons in the wide range of skills associated with mechanized infantry, even before the Bradley entered troop service. According to the Chief of Infantry, Major General Michael F. Spigelmire: "The purpose behind the restructuring was to reduce the time lost in dismounting the the fire teams, to improve the fire and maneuver capability of the squads, and to align the Bradley-equipped mechanized infantry with the rest of the force."[7]

The new organization was designed on the basis of a 1988 Infantry School white paper that addressed the doctrine, training, and force design of mechanized infantry equipped with the BFV. This study was both written and revised on the basis of comment from the field, and was presented at the Bradley Worldwide Conference held in June 1989 at Fort Benning, Georgia. Favorable reception of the paper at the conference led to further analysis of the new concept. Concurrence by the main field commanders and the commanding general of TRADOC led to promulgation of the new platoon organization for implementation in January 1990, with formal documentation in the Table of Organization and Equipment to follow.

The gist of the reorganization was that the Infantry branch had reinvented the tank platoon. Explanation of this requires some review. Bradley platoons, and M113-equipped mechanized infantry platoons before them, as well as their predecessors, had been organized on the basis of three squads and a headquarters, in common with the organization of rifle platoons and their analogs in the U.S. Army since 1920. Perpetuating this when designing the Bradley platoon had therefore doubtless seemed a straightforward course of action. Actually operating the platoon on this basis, however, was extremely difficult. By doctrine, it was necessary for both squad leaders and assistant squad leaders to be qualified Bradley commanders, along with the platoon leader, platoon sergeant, and assistant platoon sergeant. In effect, a slightly higher proportion of the personnel in a mechanized infantry platoon had armored vehicle command responsibilities than in a tank platoon. This was a demanding arrangement; it was also a clumsy one. Again, according to doctrine, the squad leader was supposed to lead the Bradley dismount team; he thus had to climb out of the turret and switch places and equipment with his assistant, then lead the dismount team into action while the assistant squad leader lost valuable time trying to orient himself for mounted combat. The squad leader, meanwhile, found himself leading a "squad" of at most five men, unable to fire and maneuver independently.

The new organization addressed these problems by reorganizing the platoon into separate vehicle and dismount forces. The Bradleys were now formally, as well as tactically, organized into two sections of two each, in a manner precisely analogous to the tank platoon. Each vehicle carried a four-man fire team, with a squad leader for each two vehicles' teams, allowing two squads identical in structure with light infantry squads to be dismounted. There was no specific headquarters vehicle; rather, the platoon headquarters and its augmentation (the artillery forward observer, his radiotelephone operator, and the platoon's attached medic) were distributed among the empty seats of the various Bradleys.

The Chief of Infantry, Major General Spigelmire, pointed out that this reorganization, which involved no net change in numbers or rank allotments, had a variety of advantages, streamlining training, providing each vehicle in the platoon with a fire team it could dismount for security missions while restoring fire and maneuver capability to the squads, and speeding the dismount process. Another advantage claimed for the organization was that it "aligns the Bradley-equipped mechanized infantry with the rest of the infantry force." Certainly it aligned the squad with the rest of the force; two-by-two organization and specialist vehicle crews, on the other hand, suggest alignment of the platoon, with the armor force.[8]

This reorganization marked a doctrinal watershed for the mechanized infantry, but it had been a long time coming. The tactical organization of the Bradley platoon into sections for mounted fire and movement under the 1986 edition of FM7-7J has been remarked on above. Similarly, Bradley gunnery, a large fraction of the platoon's training program, had evolved steadily over the years from the early Infantry School syllabus to one more and more similar to armor practice.[9]

The complementary reorganization of the dismounted element to conform with light infantry rifle squads is thus interesting. The hallmark of previous mechanized infantry organizations had been a stress on structures as analogous as possible to other infantry organizations despite radically different equipment. The 1990 mechanized infantry platoon, in contrast, was a radically different organization built of standard subunits. An interesting parallel may be drawn here with Air Force historian Richard Hallion's discussion of the idea of commonality in aircraft procurement. Hallion notes that efforts at achieving commonality between services has historically been successful when components are shared between different platforms. Trying to force a single platform to suit the needs of two users, on the other hand, has historically been fraught with difficulty.[10] If the 1990 Bradley platoon represents the dividends of component commonality, it may be argued that the Bradley in its cavalry guise represents the problematic fruits of platform commonality.

On the cavalry front, some of the misgivings about the usefulness of the M3 in the reconnaissance role were being confirmed at the National Training Center. As a result, in 1989, the Armor Center commenced tests of two new scout platoon organizations. One, a composite platoon, consisted of four Bradley

CFVs and six High Mobility Multipurpose Wheeled Vehicles (HMMWVs). The other was a homogeneous platoon of ten HMMWVs, armed with a mixture of pedestal-mounted TOW launchers and Mk 19 Mod 3 40mm automatic grenade launchers, employing TOW infrared night sights as compensation for the loss of capabilities associated with the CFV Integrated Sight Unit.[11] This experiment did not occur in a vacuum; units in the field were conducting similar tests. The 1st Battalion, 64th Armor, for instance, after its M113-equipped scouts suffered badly in a November 1986 exercise at the National Training Center, went to its next one in August 1988 with an ad hoc scout platoon of eight HMMWVs. The unit was highly satisfied with the results, as the increased number of vehicles and their reduced signatures allowed better execution of the reconnaissance mission.[12] At any event, such platoons had been present for some time, as the divisional cavalry squadrons of light divisions. These had been generally satisfactory, although their smaller establishments, like most facets of light divisional organization, caused some concerns as to their sustainability during combat operations.[13] These experiments suggested that the HMMWV was more survivable than the M3 due to its greater stealth, that the larger numbers per scout platoon it afforded allowed better coverage in reconnaissance missions while being easier to support, and that the greater firepower of the CFV did not enhance its effectiveness in the reconnaissance mission. As a result, in 1990 the Army standardized on a 10-HMMWV scout platoon for its heavy battalions.[14]

These changes applied to battalion scout platoons only, not to the divisional or regimental cavalry, where doubts continued to surface about the usefulness of M3s unsupported by tanks in the heavier cavalry roles. As a result, experimental Tables of Organization and Equipment for a new, L-series, cavalry squadron were introduced to replace the J-series tables associated with the Division 86 reorganization. The first unit to be so equipped, the 3rd Squadron, 4th ("Quarterhorse") Cavalry Regiment, of the 3rd Infantry Division came on line in December 1990. This organization reintroduced the main battle tank into the divisional cavalry squadron's platoons, with each of the three platoons of the ground troops organized in a "three-by-five" (three M1, five M3) arrangement, reminiscent of the final H-series platoon's mix of M60s and M113s. The new platoon, however, was heterogeneous, with two sections each of two Bradley CFVs and one Abrams tank, led by a headquarters section equipped with one of each, rather than the homogeneous sections of previous mixed cavalry platoons. This configuration greatly enhanced firepower and was considered especially promising for the counterreconnaissance battle. At the same time, it allowed overwatching protection for scouting forces.[15]

By the end of 1990, then, both the infantry and armored cavalry communities were undergoing large-scale organizational and doctrinal saltations, in each case making wrenching adjustments to the closer combined arms atmosphere of the time. In August 1990, meanwhile, Iraq invaded Kuwait. The Army was thus required to mobilize in the middle of its reequipment effort. This raised a variety of problems, as units being sent in-theater were hastily reequipped with A2s wherever possible; approximately 2,200 Bradleys were deployed, so this represented a considerable effort. Modifications needed for this "rollover" were carried out in the United States, at the Mainz Army depot in Germany, and even in-theater at the Saudi Arabian port of Ad Damman, with the help of one hundred volunteer civilian depot workers from stateside facilities. The achievement is all the more impressive when it is considered that the even larger M1 tank fleet was undergoing a similar rollover.[16] Between rollover efforts and transfer of prepositioned machines in Germany, 48 percent of the Bradley fleet were A2s, while another 33 percent were A1s.[17]

Other aspects of reequipment were not accomplished. For instance, all three sorts of scout platoons then in the Army establishment saw service, variously equipped with M113s, M3s, and HMMWVs. In fact, a small surplus of M1 tanks in-theater led to the 1st Infantry Division (Mechanized)'s cavalry squadron, 1-4 Cavalry being modified to an ad hoc analog of the tentative L-series organization discussed above.[18]

DESERT STORM yielded considerable information about the suitability of the Bradley for armored operations, despite its lopsided nature and the lack of opportunity for dismounted combat. To begin with, the BFV's vulnerability had been less than its detractors claimed. It was, however, still considerable, especially the unmodernized machines that missed the rollover to the A2 standard, owing to early deployment. In the first U.S. ground combat of Operation DESERT STORM, for instance, on 22 January 1991, one Bradley of the 3rd Platoon, I Troop, 3rd Armored Cavalry Regiment, was holed fifteen times by rounds from medium and heavy machine guns and "a small caliber, older model antitank weapon." The two scout-observers in the rear compartment were wounded, but the vehicle was still operative and suffered no secondary fires or explosions.[19] In this it was lucky; sympathetic detonation of ammunition contributed significantly to casualties among Bradley crews.[20] The greatest proportion of Bradley losses were due to friendly fire incidents.

The operational mobility of the Bradley appears to have been uniformly satisfactory. Bradleys had little problem in keeping up with M1 tanks. The problems of differential mobility within the combined arms team, however, were only mitigated rather than abolished, as many supporting elements were still mounted in M113-derived vehicles. As a result, Improved TOW Vehicles, Fire Support Team Vehicles (FIST-V), and M577A1 armored command posts were constant drags on the armored advances of the ground war.[21] So too were mortar carriers, maintenance and support vehicles, and armored ambulances in these roles. The lack of new M113A3 vehicles projected for many units, with their improved power trains and suspensions, amplified the problem.[22] Given the firepower-intensive nature of Alliance operations, the need to keep the fire support team up with the rest of the force was the most serious problem. As an expedient, a number of units moved as much FIST equipment—laser rangefinder-designators, extra radios, and digital message devices—as possible into Bradleys, especially where the conversion to HMMWV-mounted scout platoons made surplus M3s available.[23]

Reliability also substantially exceeded expectations, with availability rates of over 90 percent and numerous cross-country moves of one hundred to three hundred miles accomplished without major breakdowns.[24] In part, this reflected the preponderance of new vehicles,[25] but adherence to prescribed maintenance procedures was also credited as a factor.[26] This suggested that units had successfully adapted to the demands of the sophisticated Bradley systems.

The weapons fit gave fewer problems than detractors had expected, especially as regarded the range incompatibility of the TOW and Bushmaster. The Bradley's weapons were apparently even be more capable than official doctrine envisioned; the account that opens the body of the present work is one example, if exceptional. The 25mm cannon was considered particularly effective. The weapon had recently received a more potent antiarmor round using Armor Piercing Fin Stabilized Discarding Sabot (APFSDS) projectiles with higher muzzle velocity, greater sectional density, and depleted uranium penetrators. They were, in effect, small versions of the projectiles fired by the much larger 120mm gun of the M1A1 tank.

On the other hand, the Bradley's sensors and fire control arrangements were shown to be inadequate to maximize the performance of the weapons available. In particular, the TOW missile's 3,750m range turned out to be difficult to exploit without a rangefinder. In addition, the thermal sight, though especially useful in penetrating sandstorms and smoke clouds, made target identification inherently difficult due to the fuzziness of infrared imaging. The Integrated Sight Unit made a relatively poor target acquisition device for the vehicle commander, who had to rely on an optical extension to make use of the sight's limited field of view. Reliability of the coaxial machine guns also proved to be a problem. In A Troop, 4th Squadron, 7th Cavalry, the 3rd Armored Division's divisional cavalry squadron, for instance, all eighteen such weapons jammed during a fight.[27] Furthermore, training practices geared toward conserving peacetime training ammunition sometimes interfered with completion of missions. Problems such as failure to engage field fortifications too heavy to defeat with 25mm gunfire with the TOW, for instance, arose in the absence of tank or artillery support.[28]

Perhaps the most persistent criticisms specific to the Bradley concerned its human engineering. The vehicle was cramped when fully loaded with personnel and equipment, much of which was awkwardly placed. As a result, units developed a variety of field expedients, such as replacing the existing troop seating with locally fabricated ammunition lockers that doubled as bench seats.[29] Crowding had been a long-standing complaint, again exacerbated by the complex layout of the Bradley's interior, with its extensive specialized stowage for ammunition and equipment, relative to that of the M113:

> Unfortunately, three squads of nine men will not fit easily into three M-2 IFVs. Fort Benning says they will fit fine, but several National Training Center rotations by the 2nd

Armored Division and the 1st Cavalry Division, as well as numerous field exercises, have proven that the full squad with all its weapons ammunition and equipment is a tight fit in an uploaded carrier—too tight.[30]

Meanwhile, outside the vehicle, the diesel engine's exhaust tended to blow over the commander when operating with the hatch open; which is to say, most of the time.[31] This last problem was the more serious given that the same exhaust plume that blew up to discommode the Bradley commander could also give away the vehicle's position even in defilade. To cope with this drawback, a variety of extemporized exhaust extensions were fitted.[32]

Finally, the lack of "medium-weight" forces had led to vulnerability during the early stages of the military buildup in Southwest Asia. This was just the potential danger raised by Huba Wass de Czege's warning over the abandonment of the middle ground in infantry with the adoption of the Bradley, and by those cavalrymen upset by the M3's bulk. The XVIII Airborne Corps, especially the 82nd Airborne Division, which had established the first U.S. ground presence in the theater, had been a thin line indeed against putative Iraqi armored thrusts into Saudi Arabia. Even though no such thrust came, the sense of vulnerability was intense. A substantial number of proposals surfaced for forces striking a balance between the strategic mobility of the light infantry divisions and the combat power of the heavier formations, whether organized as light cavalry or as motorized infantry.[33]

After the Gulf War, a new cycle of force and equipment development was set on foot to respond to the lessons learned. Once again, the assimilation of lessons learned took place against a background of shrinking resources and demobilization.

An entirely new family of armored vehicles had been planned some time before the crisis, but the collapse of the Warsaw Pact, occurring almost simultaneously with the Gulf War, delayed and attenuated this effort, as budgetary strictures and questions about the nature of future conflicts derailed established plans. As early as the end of 1985, the Army, still perceiving a growing and evolving Warsaw Pact armored threat, had set up a task force to begin work on a successor generation of combat vehicles, to replace the current heavy force. The Armored Family of Vehicles (AFV) this task force envisioned was to be based around two common chassis: a heavy chassis in the 55–65 ton range and a medium one generally expected to be derived from the Bradley. The systems employing these hulls were planned to share as many electronic and mechanical components as possible, in hopes of reaping potential logistical and training benefits. Among the major members of the family were a main battle tank and infantry fighting vehicle based on the projected heavy chassis, with identical levels of protection.[34]

By the time of Operation DESERT STORM, the basic features of the Future Infantry Fighting Vehicle (FIFV) had been established. They were an interesting mix. At first glance, the basic layout was a throwback: it was to incorporate the same two-man crew and nine-man dismount element as the old XM723, although the problems inherent in this configuration were expected to be alleviated by an advanced fire control system featuring automatic cueing and target acquisition. This was expected to allow the tactical functions of the Bradley commander and gunner to be combined, the more so because the vehicle was expected to incorporate embedded training systems. The weapons fit, on the other hand, was to be generally analogous to that of the Bradley: a rapid-fire gun of heavier but unspecified caliber and advanced antitank guided missiles. These weapons were to be augmented by a directed energy weapon intended to damage the optics of vehicle, missile, and infantry sensors and an "area suppression weapon," also of undefined characteristics. Enhanced passive protection was to be provided by M1-style modular armor arrays and compartmentation. Mobility was to be enhanced through advanced integrated propulsion and an advanced track, again in common with those of the projected "Block III" main battle tank. The FIFV was expected to reach units by the third quarter of Fiscal Year 2004.[35]

Even as the new 55-ton vehicle was being announced to the infantry community, however, it was cancelled. Congress, wary of the expense and ambition of the Army's Heavy Forces Modernization (HFM) effort to implement the AFV, had long resisted the program. The collapse of the Warsaw Pact, together with the encouraging performance of the Abrams and Bradley in the Gulf, threw the program into further doubt. On 30 October 1991, the Army submitted a position paper to Congress restructuring the program, shelving the Block III tank and FIFV.[36] On the other hand, a separate program for a Future Scout Vehicle (later Future Scout Combat System), to replace the M3 and HMMWV, has survived to date.

The challenge now was to apply the technological advances of the AFV initiative and the lessons learned in Southwest Asia to existing equipment. Pursuant to this goal, the Army Infantry Center began consideration of a third block of improvements to the BFV. By 1993, a variety of upgrade plans had been put forth using the A2 version of the Bradley, with its enhanced powertrain and protection, as a springboard.[37] The new versions would be externally and mechanically almost indistinguishable from the old, but would incorporate extensive electronic improvements. These included a laser rangefinder, Global Positioning System receiver for improved navigation, and provision for a driver's thermal viewer. This last item had been successfully tested on the Bradley during the war by a troop of the 24th Infantry Division (Mechanized)'s divisional cavalry squadron (A/2-4 Cav), which found the DTV to be a great improvement over the AN/VVS-2 light-amplifying viewer normally fitted.[38] Internally, further restowage was planned, while the dismount element, now the fire team-plus-supernumeraries of the 1990 "two-by-two" platoon organization, traded their individual seats for face-to-face bench seats. Elimination of the need to serve the lateral Firing Port Weapons opened up valuable room.[39] Of the 6,724-vehicle Bradley fleet, 1,433 Bradleys were to be rebuilt to this standard, called M2A2 ODS, for "Operation DESERT STORM." An additional 1,602 were to be further upgraded to an even more advanced A3 standard, while the remainder, 2,005 unmodified vehicles in the training base and lower-readiness units and 1,684 A2 models, were to await authorization for further upgrades.[40] The A3 version was to add a commander's independent thermal viewer and digital fire control. This would allow much more rapid engagements, for the commander could now acquire targets independently and automatically hand them off to the gunner. In addition, a digital databus would be added to network all this equipment with the help of the Intervehicular Information System (IVIS) battle-management device. The vehicle was also to be fitted for, but not necessarily with, advanced appliqué armors. All this was projected to cost $1.1 million per vehicle for those already in the A2 configuration, or $1.8 million for earlier machines.[41] In short, the A3 Bradley was to be equipped equivalently with, but not identically to, the M1A2 main battle tank it was to complement, precisely the situation which the AFV initiative's emphasis on commonality was intended to avoid, though some components at least were common.

Nor was the postwar modernization plan limited to qualitative improvements. In much the same way as the Black Death raised the per capita income of late medieval Europeans, a side effect of the drawdown in the wake of the Persian Gulf cease-fire and the lessening of East-West tensions, together with the replacement of many M3-equipped scout platoons by HMMWV organizations was, in effect, a surplus of Bradley chassis relative to the force structure. As a result, Bradleys and their derivative FVSCs, the multipurpose vehicle developed to carry the Multiple Launch Rocket System, became available for a variety of support roles. The former were developed into fire support vehicles for air and artillery forward observer teams with laser rangefinder/designator packages replacing the TOW missile launcher and as interim forward air defense vehicles carrying Stinger man-portable antiaircraft missiles under armor in the wake of the collapse of the Army's Forward Area Air Defense plan, as well as being projected to replace M901 Improved TOW vehicles in mechanized infantry Echo (antiarmor) companies.[42] The latter were planned for use as a basis for new command, armored ambulance, and electronic warfare vehicles.

In addition, a 1993 revision of *FM 7-7J, Mechanized Infantry Platoon and Squad (Bradley)* was issued to reflect DESERT STORM experience. Its most notable innovation was a reorganization of the squad, where, once again, the goal of commonality with other infantry small-unit organization was abandoned to enhance firepower. In this case, the two identically-equipped fire teams of the previous organization were modified, one gaining an additional M249 Squad Automatic Weapon, the other gaining an additional M203 grenade launcher.[43]

To summarize, operational experience with the Bradley led to both additional hardware improvements as mentioned above, and new organization and doctrine. In particular, the platoon organization was radically revised to a form parallel to the tank platoon rather than the traditional infantry rifle platoon, and the concept of mounted combat for the infantry was radically revised. At the same time, cavalry units were deemphasizing the CFV, as lighter vehicles replaced it in battalion scout platoons and it was supplemented by main battle tanks in divisional and regimental cavalry squadrons. Operation DESERT STORM transpired in the midst of this reorganization and reequipment, providing the Bradley's only combat experience to date. This was generally successful, although room for improvement was found in the areas of protection and fire control equipment. At the time of writing, such improvements are planned for incorporation in the near future, given that a much larger and more tanklike successor, the FIFV was cancelled in the wake of the collapse of the Warsaw Pact.

NOTES

1. Peter M. McVey, *Sending the Very Best: An Oral History Interview with Major General Peter M McVey, Program Executive Officer, Armored Systems Modernization* (Warren, MI: U.S. Army Tank-Automotive Command, Historical Office, 1993), 26.

2. Michael Green, "Better Bradley," *International Combat Arms* 7 (January 1989): 57f.

3. The swim curtain was a canvas barrier providing buoyancy sufficient for the Bradley to cross modest water obstacles, using its tracks for propulsion. A defect in part of the system used to collapse the barrier upon emergence from the water led to a number of swamping incidents. Although the total number of such incidents was small—eleven incidents in approximately ten thousand uses of the flotation system—it led to further journalistic attention and congressional hearings just as the Burton affair was winding down. Its major significance to this study is a further index of public and Army confusion over the rationale for the infantry fighting vehicle; see below. See U.S. Congress. House of Representatives. Armed Services Committee. Subcommittees on Procurement and Military Nuclear Systems, and Investigations, *Swim Capability of the Bradley Fighting Vehicle* (Washington, DC: 100th Congress, 1st Session; 1988) for an overview.

4. Army Deputy Chief of Staff for Operations.

5. McVey, *Sending the Very Best*, 36f.

6. A new model of the M113 had entered production in February 1987, featuring power train upgrades in addition to these improvements; "Infantry News," *Infantry* 75 (September–October 1985): 8.

7. Michael F. Spigelmire, "Bradley Platoon Organization," *Infantry* 80 (January–February 1990): 1f.

8. *Ibid.* Cf. a similar proposal by an armor officer at the Armor School's Directorate of Combat Developments to split the Bradley equipped mechanized infantry company into two rifle platoons identical in structure to the those of light infantry units, and two Bradley platoons identical to the standard six-Bradley scout platoon: Chester A. Kojro, "Bradley Platoon Organization," *Infantry* 77 (March–April 1987): 16–18. This idea has exhibited staying power: over a decade later, the July–December 1997 issue of the same journal carried an article, "Let's Reorganize Our BFV Companies," by Major Steve E. Landis, a company trainer at the National Training Center, repeating the same idea without reference to the earlier work.

9. Thomas T. Smith, "Bradley Gunnery Evolution," *Infantry* 78 (May–June 1988): 25. Captain Smith was originally a junior officer in the 1st Battalion, 41st Infantry, the first unit to receive the Bradley, and remained with the unit for six years, eventually becoming a company commander.

10. Richard P. Hallion, *The Historical Evolution of Commonality in Fighter and Attack Aircraft Airframe Development and Usage* (Edwards AFB: History Office, Air Force Flight Test Center, 1985), 72–76.

11. "New Scout Platoon Concept Will Test HMMWVs as 'Stealthy' Scouts," *Armor* 98 (March–April 1989): 51.

12. Barry Scribner, "HMMWVs and Scouts: Do They Mix?" *Armor* 98 (July–August 1989): 33. Major Scribner was operations officer of the 1st Battalion, 64th Armor at the time.

13. Jeff Witsken and Lee MacTaggart, "Light Cavalry in the 10th Mountain Division," *Armor* 99 (July–August 1990): 36–40. Captain Witsken was operations officer of the 10th Mountain Division's cavalry squadron, 3-17 Cavalry; Captain MacTaggart was assistant S-3 of the 10th's National Guard roundout brigade.

14. John C. Woznick, "The Scout Vehicle," *Armor* 103 (September–October 1994): 32. Lieutenant Colonel Woznick had recently served as Cavalry branch chief, Maneuver Division, Directory of Combat Developments at the Armor Center, Ft. Knox, Kentucky.

15. George J. Salerno, "Repairing the Broken Sabre: Overview of L-Series Divisional Cavalry," *Armor* 103 (January–February 1994): 29. Captain Salerno had been a junior officer in the Quarterhorse when it converted, and was currently commander of C Troop, 2-4 Cavalry, 24th Infantry Division. This experimental organization was shelved in favor of one paralleling regimental armored cavalry troops: two M3-equipped scout platoons, two M1-equipped tank platoons, and a section of two heavy mortars.

16. McVey, *Sending the Very Best,* 41.

17. U.S. Department of Defense, *Conduct of the Persian Gulf War* (Washington, DC: USGPO, 1992), 739.

18. Robert Wilson, "Tanks in the Divisional Cavalry Squadron," *Armor* 101 (July–August 1992): 6. Lieutenant Colonel Wilson was commanding officer of 1-4 Cavalry during Operation DESERT STORM.

19. Jonathan J. Negin, "DESERT STORM—The First Firefight," *Armor* 94 (March–April 1994): 8. CPT Negin was the leader of the platoon.

20. John R. Sutherland III, "Innovations from Operation DESERT STORM," *Infantry* 85 (May–June 1995): 16. Captain Sutherland was a mechanized infantry company commander in the 24th Infantry Division (Mechanized) during DESERT STORM.

21. *Conduct*, 739.

22. A. A. Puryear and Gerald R. Haywood II, "Ar Rumalyah Airfield Succumbs to Hasty Attack," *Armor* 100 (September–October 1991): 19. Puryear and Haywood were a troop executive officer and scout platoon leader, respectively, of the 2nd Squadron, Third Armored Cavalry Regiment, during DESERT STORM.

23. Sutherland, "Innovations," 17f. Another application for these machines was as mounts for mechanized infantry executive officers not previously allotted Bradleys, despite their prescribed role as combat leaders (see chapter 8).

24. *Ibid.*

25. McVey, *Sending the Very Best,* 43.

26. *Conduct*, 739.

27. Martin L. Shupe, "M1A1, Bradley Defied the Critics," *Armor* 100 (May–June 1991): 12. Sergeant Major Shupe was with the VII Corps Public Affairs Office.

28. Negin, "DESERT STORM—The First Firefight," 9.

29. Sutherland, "Innovations," 16.

30. Scholtes, "Where Have All the Infantrymen Gone?" 92.

31. *Conduct*, 740.

32. Sutherland, "Innovations," 17.

33. The debate over strategic mobility versus combat power in the Army force structure is at least as old as U.S. rearmament for World War II, and can be tracked through the 1970s in the relevant chapters of Captain Jonathan M. House's *Toward Combined Arms Warfare*. This old controversy flared up again with the design and activation of the light infantry divisions, seen by many analysts as having sacrificed too much combat power and operational mobility in fulfillment of an essentially arbitrary strategic mobility parameter—the upper limit of five hundred sorties by C-141 airlifters—and continued up until DESERT STORM. Michael J. Mazarr, *Light Forces and the Future of U.S. Strategy* (New York: Brassey's, 1990), is a useful if partisan summation. Mazarr continued in this vein in the light of Southwest Asian experience in his article, "Middleweight Forces for Contingency Operations," *Military Review* 71 (August 1991): 32–37; as mentioned, this was the first of a spate of calls for such forces in Army professional journals. Nor was this strictly an intellectual exercise: in April 1993, the Army converted its 3rd Armored Cavalry Regiment to light cavalry, mounted in HMMWVs as a covering force for XVIII Airborne Corps (George F. Oliver III, "Light Cavalry Has Strategic Value Today," *Military Review* 74 [November 1993]: 65f. Lieutenant Colonel Oliver commanded the 1st Squadron, 3rd ACR at the time.). It also approved acquisition of the XM8 Armored Gun System—another euphemism for "light tank" in the Army lexicon (cf. the Sheridan "Armored Reconnaissance/Airborne Assault Vehicle")—an item that had been pursued on and off for over a decade and a half. The XM8, in turn, was cancelled in early 1996.

34. Eric Ludvigsen, "The Mechanized Force in the Next Century," *Army* (July 1987): 23f. In May 1990, the program was renamed Armored Systems Modernization.

35. "Chief of Infantry Update" ("Infantry News" column), *Infantry* 81 (July–August 1991): 8.

36. McVey, *Sending the Very Best*, 53f.

37. *Ibid.*, 33.

38. David L. Gallop, "The Driver's Thermal Viewer: A DESERT STORM AAR," *Armor* 103 (May–June 1994): 17. Captain Gallop commanded A/2-4 Cav in DESERT STORM.

39. Bernard Adelsberger, "Bradley to Get Better in Stages," *Army Times*, 25 May 1993, 20.

40. Jerry A. White, "Bradley Modernization: Making a Great System Better," *Infantry* 83 (November–December 1993): 1–2. Major General White was Chief of Infantry.

41. John Boatman, "A3 Standard Studied for Bradley Fleet," *Jane's Defense Weekly* 16 (January 1993): 8.

42. White, "Bradley Modernization," 2.

43. U.S. Army. Headquarters, Department of the Army. *Mechanized Infantry Platoon and Squad (Bradley)* (Washington, DC: Headquarters, Department of the Army, 1993), Appendix A. The Army Warfighting Experiments held in the wake of the Gulf War led to a recommendation that a five-man machine gun squad be added to this organization; later, the "Division XXI" reorganization announced in the summer of 1998 specified three dismount squads of nine in the four-vehicle mechanized infantry platoon, while reducing the battalion to three companies. How the extra squad will be accommodated is unclear as this book goes to press. It may simply be a belated but radical recognition that allotted manpower is always less than battle strength.

Conclusion

> The search for new "concepts" has been underway for 20 years, and nothing of great consequence has appeared. This is not because there are no new concepts available. The real problem lies with those who are hunting for something new. They want a new look, but one that is familiar enough to suit their own tastes.
>
> —Colonel Robert P. Zeigler, USA, Chief
> Organization and Doctrine Division,
> Headquarters, Combined Arms Group,
> Fort Leavenworth, July 1966[!][1]

On the foregoing evidence, it would appear that the U.S. Army experienced great difficulty in fielding an infantry fighting vehicle because it set out to implement a set of radical doctrinal shifts by conservative means. The Army proposed to change the role of the infantry armored vehicle from transport to combat while expanding the role of the troops they carried; at the same time, it proposed to do so by evolving the new one from the old, although the characteristics of the two diverged sharply. These contradictions resonated with the Army's elephantine acquisition system and the general lack of resources for modernization to turn the Army's adoption of an IFV into two decades of impasses, controversies, and complications.

The Army's original armored infantry had been a specialized element of the Armored Force. While its weapons and tactics were fairly conservatively derived from those of the regular infantry, it was nonetheless a force specifically oriented toward the close support of armored combat commands at the tactical level, never existing as units larger than a battalion even on paper, and—through the process of cross-attachment into armored task forces and smaller combined arms combat teams—usually in even smaller organizations than that. Although an armored infantry division was proposed by the Armored Force early in World

War II, it was never fielded due to a lack of utility.[2] Mechanized infantry was a different proposition, a generalist force under the proponency of the Infantry branch, using organic armored vehicles to carry out the whole spectrum of infantry missions, organized into formations of up to division size.

The radical change wrought by the idea of mechanized infantry, however, was implemented conservatively; the new mechanized infantry continued the organization of earlier armored infantry units. At the same time, however, the armored infantry mission was subsumed by mechanized infantry, replacing a specialized force under the proponency of the Armor branch with a generalist one under that of Infantry. For an army that took the benefits of standardization as an article of faith, one further haunted by the specter of mobilization, such doctrinal monism had undeniable appeal.

As with training and doctrine, so with materiel. In setting the requirements for the new infantry vehicle, the Army's planners used the existing armored personnel carrier as a point of departure. From one point of view, this made sense: after four generations of design, such vehicles were well-understood and technically mature. They were relatively simple and inexpensive, capable of carrying an infantry rifle squad and its equipment over challenging terrain, even across modest water obstacles, and themselves capable of being transported by air even in medium-sized airlifters such as the Lockheed C-130.

Unfortunately, these qualities, useful as they were, warred at almost every point with the increased protection and firepower necessary for an infantry fighting vehicle to take a direct part in the armored battle. Attempts to reconcile this fundamental conflict led to the loss of a decade's time for modernization as the service and the Department of Defense rejected one vehicle after another in pursuit of an unattainable ideal. When the Army finally got an IFV, the Bradley, it did so primarily by relaxing its original, APC-derived, requirements. By this time, however, it was necessary to keep many earlier elements of the previous design to save time and money, despite the need to add features to accommodate the new (and hitherto nonexistent) role of cavalry fighting vehicle. Many characteristics of the Bradley were thus set in stone, or rather in 5083 aluminum, before much gain could be made from trade-offs. Had the requirement to carry a squad-sized dismounting element been relaxed earlier, for instance, the result would have been a smaller and more portable vehicle that might have been more credible in the armored cavalry role; had the airportability requirements been relaxed earlier, a more capacious and/or better protected vehicle—possibly posessing extensive commonality with its contemporary main battle tank, with consequent tactical, training, and logistical advantages—might have resulted. Whether a smaller or larger vehicle was pursued, earlier relaxation of the requirement for swim capability would have led to a simpler and better protected vehicle in the end.

When the Bradley entered service, a conservative approach to training and doctrine again became problematic. It was just possible to regard an M113-equipped mechanized infantry platoon as a rifle platoon that happened to have four armored vehicles in its inventory; attempting to drop the more capable and demanding Bradley into the same was distinctly problematic, yet this was in many ways the case. The demands and capabilities of the Bradley led to concern about the decay of traditional dismounted infantry skills. Although this study early on dismissed much of the relevance of dragoons and mounted infantry as precursors, it should be noted that their experience is analogous here: almost all dragoons suffered functional drift, as accommodating the demands and capabilities of the horse led to compromises of dismounted combat skills and combat power.[3] This phenomenon continues: even in 1995, a Command and General Staff College paper dealt with the problem of "Why Johnny Can't Dismount."[4] Steps have been taken to ameliorate this problem somewhat; once again, it has been dealt with largely by relaxing the demand for conformity with more traditional infantry practice. The Army's recently announced Division XXI reorganization appears to retreat from this trend; whether it will be tenable remains to be seen.

From the foregoing, a paradoxical and somewhat dispiriting conclusion may be drawn: it would appear that the infantry fighting vehicle has been problematic for the U.S. Army to the extent that the infantry has been involved in it. The vision of a mechanized infantry force wedded to armored fighting vehicles yet preserving the general-pupose nature of the earlier force repeatedly led the Army to pursue unrealistic technological goals and forced it to make awkward trade-offs when they proved unattainable. These trade-offs, embodied in the Bradley Fighting Vehicle, in turn led to problems with training and support, distorted force designs, and a doctrine that was at best almost completely incomprehensible to outsiders and at worst almost completely incomprehensible to those within. To the extent to which the Bradley has been successful, success has been achieved through incrementally moving away from this vision, and most importantly through the efforts of its users to master and exploit its capabilities than through planners' efforts to define them into institutional tractability.

Could these problems have been resolved had the Armor branch retained proponency for armored infantry, or at least for its vehicular elements, during the ROAD reorganizations around 1960? It is hard to say, but Armor branch proponency might have lowered some technological and logistical hurdles by facilitating greater harmony or even commonality between the design of main battle tank and infantry fighting vehicle. On the doctrinal front, some of the conflicts inherent in organizations trying to operate armored fighting vehicles within the framework of the infantry ethos and mission might have been avoided by seeing the dismounting element as a specialist subset of an armored force rather than seeing armored fighting vehicles as an element of equipment within a general-purpose infantry force element. More practically, it appears reasonable to suppose that a heavy force whose team commanders, for example, had led both tank and mechanized infantry platoons as part of a normal career path might train more smoothly and fight more capably. Were such teams—call them "armor companies," perhaps built around two tank and two mechanized infantry platoons, subject to tailoring by cross-attachment as now[5]—the norm in garrison as well as in the field, they might function more effectively. Certainly the Armor branch has experienced an analogous situation in operating its armored cavalry units, with their specialized mission composite organization and dismountable elements. To reach further back, a rare viable example of a mounted force successfully combining mounted mobility and largely dismounted combat is available in the history of the U.S. Cavalry branch in the Civil War and the later operations on the Great Plains.

Failing the achievement of such increased integration, dividing the vehicular and dismounted parts of the mechanized infantry between the two branches—that is, assigning the operation if the IFVs themselves to the Armor branch—might also have paid off in simpler training and administration.[6] It is interesting to note that, like the mechanized infantry, all but one subbranch of the Army's infantry force (as well as the United States Marine Corps) define themselves partly by their means of transport, but these means are not organic and thus place fewer demands on the training syllabus and institutional culture of their units.

Such shifts in proponency could be valid even today; they would be radical, but certainly in this case, whenever it has taken the path of least resistance institutionally and bureaucratically, the Army has ended up taking the path of greatest resistance doctrinally and technologically.

POSTSCRIPT

In late 1998, the United States Army announced that it was drawing up plans for a Future Infantry Vehicle (FIV), projected to enter service in 2012. The FIV initiative called for a vehicle with a limited amphibious capability, armed with a medium caliber cannon and a missile system, with a crew of two or three plus a full rifle squad; in order to permit carriage in a C-130J medium airlifter, the service set a weight limit of 22.5 tons.[7]

NOTES

1. Robert P. Zeigler, "Mechanized Fighting Vehicle," *Military Review* 46 (July 1966): 3.

2. Kent Roberts Greenfield, Robert R. Palmer, and Bell I. Wiley, *The Organization of Ground Combat Troops* (Washington, DC: USGPO, 1947), 324f.

3. Two recent studies of the capabilities and limitations of nominal cavalry forces operating as mounted infantry can be found in Volume IV of the Marquess of Anglesey, *A History of the British Cavalry, Volume 4: 1899–1913* (London: Leo Cooper, 1986) dealing with the Boer War and its aftermath; and David Evans, *Sherman's Horsemen: Union Cavalry Operations in the Atlanta Campaign* (Bloomington and Indianapolis: Indiana University Press, 1996).

4. Edward G. Gibbons, Jr., *Why Johnny Can't Dismount: The Decline of America's Mechanized Infantry Force* (Ft. Leavenworth, KS: U.S. Army Command and General Staff College, 1995).

5. It is worth noting that this balanced company team organization—augmented by an armored engineer platoon and the necessary logistics package—is the usual heavy-force augmentation to light infantry brigades training at the Joint Readiness Training Center, Fort Polk, Louisiana; JRTC is the light-force analog of the National Training Center discussed above. See Paul E. Thompson, Jr., "Light/Heavy Integration at the Joint Readiness Training Center," *Armor*, July–August 1998, 10–14. Sergeant First Class Thompson wrote his article as a JRTC observer-controller. The term "armor company" derives from Richard Simpkin's use of the term "armor platoon" for the basic subunit of MBTs and IFVs he proposes in his book *Mechanized Infantry* and elsewhere. Simpkin's organizational concept for the main armored force appears to have had some affinities with U.S. Army armored cavalry practice.

6. This was tried as part of the Pentomic reorganization of the infantry division and generally considered unsuccessful; however, this was in a different operational context. In any event, it is tempting to consider any Pentomic experience as an unfortunate special case due to the nuclear-oriented doctrinal baggage associated with it; see A. J. Bacevich, *The Pentomic Era*, for details.

7. "US Army Studies FIV to Replace Bradley," *International Defense Review*, October 1998, 21.

Bibliography

SOME NOTES ON THEORY AND HISTORIOGRAPHY

A large part of the difficulty of studying mechanized infantry is that the subject is almost invisible to mainstream military history, falling as it does between infantry and armor. Each of these branches has a large and rich literature, as does the practice of force design. Each, however, tends to leave the infantry component of armored forces to the others. This is true to a considerable extent even with such comprehensive works as John English's *On Infantry*.[1]

"Separated from their infantry support, the —— tanks were defeated in detail" is as much a stock phrase as Robert O'Connell's lampoon of the dismissive treatment given the study of weapons by many scholars: " 'Thus armed with iron the Hittites drove the bronze-clad Egyptians from the field,' and the like." The doctrinal and technological efforts made to avoid these calamities, however, tend to remain obscure.[2] Some of this obscurity is inherent in the nature of mechanized infantry, which closes off the traditional avenues of historical self-promotion. As discussed above, such forces do not fit in well with branch-oriented history. At the same time, the nature of their employment also tends to foreclose the option of unit history. As one mechanized infantry officer explained to an audience of his colleagues:

> Armored infantry, as units, and their armored counterparts have a strictly transitory life. Although they exist in garrison for administrative purposes, they die the moment the first shot is fired, when suddenly they are transformed into company teams and battalion task forces composed not of infantry *or* armor, but of infantry *and* armor.[3]

To say the least, this phenomenon makes it difficult for students of the field to build on existing secondary literature. There are, however, some analogous studies, although the limits of analogy must be acknowleged. The canonical place to begin discussion of troubled weapons systems is the literature on the F-111 program, in particular, Robert F. Coulam, *Illusions of Choice: The F-111 and the Problem of Weapons Acquisitions Reform*[4] and Robert J. Art, *The TFX Decision: McNamara and the Military.*[5] As with the Bradley, the F-111 shows the problems inherent in the attempt to adapt a single vehicle to two disparate missions, especially when the missions themselves are not well-defined. Coulam's work is particularly valuable for its application of John D. Steinbruners's cybernetic paradigm of decision making[6] to defense procurement issues. This model of organizational behavior, stating that actors in bureaucracies work to minimize short-term uncertainty rather than acting in pursuit of big-picture goals as postulated by the competing "analytic paradigm," certainly bears on the Bradley case. A more recent aerospace analog of the Bradley controversy, in addition, may be found in James P. Stevenson, *The Pentagon Paradox: The Development of the F-18 Hornet.*[7] Stevenson, a defense journalist and editor, sets forth one more example of a system adapted to a dual role with problematic consequences. In this case—as with the Bradley—the system reconciles the needs of two communities within one service rather than two services. His account is heavily grounded in the viewpoint of the defense reform movement, with the accompanying strengths and weaknesses.

Nor is aerospace history the only source of analogous studies. Charles Baily's *Faint Praise: American Tanks and Tank Destroyers during World War II*[8] provides insight into the problems of weapons systems that fall outside established arms of a single service. It also discusses at length the origins of U.S. Army antitank doctrine and practice, an important aspect of infantry fighting vehicle development. However, many of the vicissitudes of the Tank Destroyer Command in World War II derive from its lack of affiliation to the established arms of the service; those of the Bradley derive from the opposite situation. Edward C. Ezell, in *The Great Rifle Controversy: The Search for the Ultimate Infantry Weapon, from World War II and Vietnam and Beyond,*[9] provides another useful study, showing the tensions between the acquisition of new capabilities while preserving old ones in its treatment of the M16 rifle program. It also shows the tensions involved in competition between an outside system available immediately and an ideal one still on the drawing board, a vital matter for the progress of the Bradley's precursors in the 1960s.

In addition to these studies, the student is aided by the existence of a certain amount of work on armored warfare theory and practice that transcends the jargon of official publications and the superficialities of "buff" literature. Two contributors to this body of work stand out. Richard M. Ogorkiewicz's expositions of armored vehicle technology go far toward familiarizing the naive reader with this complex subject. Similarly, the works of the late Brigadier (Retired) Richard E. Simpkin, British tanker, vehicle developer, and technical translator, provide a useful comparative perspective on tactics and doctrine.

Traditional primary literature is also problematic. Archival access to records dealing with weapons systems still in production is almost a contradiction in terms; even older systems are recent enough that work on them is badly encumbered by security restrictions, remote and dispersed locations, and large groups of records in dead storage but not yet archived. A large body of public documents—in particular, Congressional ones—exists, but these deal mainly with the procurement process. To indulge in one more use of Willie Sutton's phrase, "that's where the money is."

Similarly, procurement issues ultimately dominate the mass of partisan literature associated with the Bradley controversy of the mid-1980s. Where it does not, it presents a strange mixture of valid, incisive critique and facile polemics requiring separate treatment (see chapter 10). This literature is part of an interesting problem in the history of ideas, but its relation to the idea under study here is often peripheral.

A knowledge of the procurement aspects of the Bradley Fighting Vehicle and its precursors is necessary, but not sufficient. Any number of weapons systems went through the process, and the bulk of them failed to do so quickly or gracefully. However, it is hard to find an example of one that had such a difficult time of it for so long a time without cancellation, and without an end to its problems once in service. The one extended scholarly work on the infantry fighting vehicle in the United States in the open literature, Daniel Joseph Kaufman's "Organizations, Technology, and Weapons Acquisition: The Development of the Infantry Fighting Vehicle,"[10] shares this bias toward procurement. As such, his work is invaluable in following the Bradley and its progenitors through the complex and exhausting process, but also as such, its conclusions, which concern themselves with adaptation of the procurement process in general to new technological realities, are unrewarding to the student of other aspects of military institutions.[11]

The requirements for weapons systems—in the informal sense—ultimately come from outside the procurement system. They are rooted in the attitudes and outlooks of service members. A useful source for exploring these mentalities in the U.S. Army is the corpus of professional military journals, which are extensively quoted in this work. As A. J. Bacevich states in his book *The Pentomic Era*:

> The emphasis on Service journals does not reflect a belief that the written musings of relatively junior officers influence American military policy to any degree. They do not. While the institutional organs of other professions presage and often inspire new developments, American military journals tend instead to reflect ideas that already enjoy official sanction. They mirror American military thought rather than determine its direction. Although the placid character of American military journals minimizes their utility as a forum for debating new ideas, this character makes them ideal for the historian attempting to understand the mind-set of the officer corps at a particular time.[12]

The following bibliography is divided into unpublished and published sources.

Because of the nature of this work, no attempt has been made to divide the latter into primary and secondary sources in order to avoid pointless anguish as to whether, for example, a retired armor officer's discussion of armored-warfare theory in a nonservice journal qualifies as "primary" or "secondary" literature.

Over and above consultation of the standard general bibliographic resources, the bulk of the published material derives from an extensive longitudinal study of *Armor*, *Infantry*, and *Military Review* publication runs since World War II combined with a search of the *Air University Library Index of Military Periodicals* for the same era. Considerable material in excess of actual citations is given here, in hopes of presenting an overview of mechanized infantry issues since 1945.

NOTES

1. New York: Praeger, 1981, 1984; the significantly different revised version, with Bruce I. Gudmundsson (Westport, CT: Praeger, 1994) shares this problem.

2. Robert L. O'Connell, "Putting Weapons in Perspective," *Armed Forces and Society* 9 (spring 1983): 444.

3. Hovell, "Armored Infantry Is Different," 49.

4. Princeton: Princeton University Press, 1977.

5. Boston: Little, Brown, 1968.

6. John D. Steinbruner, *The Cybernetic Theory of Decision: New Dimensions of Political Analysis* (Princeton: Princeton University Press, 1974), explicated in Coulam, chapter 1.

7. Annapolis: U.S. Naval Institute Press, 1993.

8. Hamden, CT: Archon, 1983.

9. Harrisburg, PA: Stackpole, 1984.

10. Unpublished diss., Department of Political Science, MIT, Cambridge, MA, 1983.

11. Kaufman concludes that the Bradley case shows that organizational complexity makes weapons development an unavoidably lengthy process, which should be coped with by accepting lengthy development times, and coping with them through extensive use of product improvement packages applied to extant systems (Kaufman, 516).

12. A. J. Bacevich, *The Pentomic Era: The U.S. Army between Korea and Vietnam* (Washington DC: National Defense University Press, 1986), 5.

UNPUBLISHED SOURCES

Kaufman, Daniel Joseph. "Organizations, Technology, and Weapons Acquisition: The Development of the Infantry Fighting Vehicle." Unpublished diss., Department of Political Science, MIT, Cambridge, MA, 1983.

Starry, Donn A. Papers, Personal Correspondence Files. United States Army Military History Institute, Carlisle Barracks, PA.

United States. Department of the Army. Combat Development Command. Headquarters. "CDC Position, MICV IPR 5–6 June 69" (memo, 4 June 1969, from "CDCMR-W"). National Archives and Records Administration, Federal Records Center, Suitland, MD, Box 338-90-512, "PM-FVS PL & ANAL 1969."

United States. Department of the Army. Project Manager, Mechanized Infantry Combat Vehicles. "Austere MICV" (brochure, corrected proof). National Archives and Records Administration, Federal Records Center, Suitland, MD, Box 338-90-512, "PM-FVS PL & ANAL 1969."

United States. Department of the Army. Project Manager, Mechanized Infantry Combat Vehicles. "MICV–XM723 System Development Plan Pre-IPR, 2 April 1969" (memo). National Archives and Records Administration, Federal Records Center, Suitland, MD, Box 338-90-512, "PM-FVS PL & ANAL 1969."

United States. Department of the Army. Project Manager, Mechanized Infantry Combat Vehicles. "MICV—XM723 Technical Performance Narrative" [1 December 1968]. National Archives and Records Administration, Federal Records Center, Suitland, MD, Box 338-90-512, "PM-FVS PL & ANAL 1969."

United States. Department of the Army. Project Manager, Mechanized Infantry Combat Vehicles. "Minutes of Systems Status Evaluation (SSE) for Mechanized Infantry Combat Vehicles, 31 July 1969" (memo). National Archives and Records Administration, Federal Records Center, Suitland, MD, Box 338-90-512, "PM-FVS PL & ANAL 1969."

United States. Department of the Army. Project Manager, Mechanized Infantry Combat Vehicles. "Project Master Plan: Mechanized Infantry Combat Vehicle, MICV—XM723." National Archives and Records Administration, Federal Records Center, Suitland, MD, Box 338-90-512, "PM-FVS PL & ANAL 1969."

United States. Departmernt of the Army. Tank-Automotive Command. Program Manager, Mechanized Infantry Combat Vehicles. "MICV Systems Messages." United States Army Tank-Automotive Command Archives, Warren, MI.

PUBLISHED SOURCES

Adams, John A. "Heavy versus Light Forces: A Middle Ground." *Military Review* 66 (October 1986): 64–73.

Adan, Avraham (Bren). *On the Banks of the Suez: An Israeli General's Personal Account of the Yom Kippur War*. Novato, CA: Presidio, 1980.

Adelsberger, Bernard. "Bradley to Get Better in Stages." *Army Times*, 25 May 1993, 20.

Aitken, Lowell A. "The Panzer Grenadiers Roll Again." *Army* 14 (February 1964): 37–40.

Aitken, Lowell A. "West German Multi-Purpose Equipment." *Infantry* 54 (September–December 1964): 58–60.

Allen, Matthew. *Military Helicopter Doctrines of the Major Powers, 1945–1992: Making Decisions about Air-Land Warfare*. Westport, CT: Greenwood, 1993.

Alter, Jonathan, and Phil Keisling. "35 Ways to Cut the Defense Budget." *Washington Monthly*, April 1982, 12–38.

Altomare, Nicholas F. "Bradley NET." *Infantry* 73 (July–August 1983): 29–30.

Anderson, Roland B. "New Weapons for the Army." *Ordnance* 51 (September–October 1966): 154–58.

Andreacchio, Nicholas A. "The Armor Battalion of the Heavy Division." *Armor* 86 (November–December 1977): 18–20.

Andress, Henry C. "The Bradley Challenge." *Infantry* 81 (January–February 1991): 18–21.

Andrews, L. G. "Chariot for the Queen." *Infantry School Quarterly* 46 (October 1956):

78–88.
Andrews, Michael A. "Active Defense Cavalrymen." *Armor* 88 (July–August 1979): 20–22.
Andrews, Michael A. "An Organization for Combined Arms." *Armor* 88 (May–June 1979): 11.
Anglesey, Marquess of. *A History of British Cavalry, Volume 4: 1899–1913*. London: Leo Cooper, 1986.
Antal, John F. "Unit Battle Drills." *Infantry* 73 (July–August 1983): 31–32.
Appling, Leo A., Jr. "APCs of the Past" (letter). *Infantry* 72 (January–February 1982): 49.
"Armored Infantry Battalion Organization." *Armor* 92 (March–April 1983): 61.
Armstrong, David A. *Bullets and Bureaucrats: The Machine Gun and the United States Army, 1861–1916*. Westport, CT: Greenwood, 1982.
Armstrong, Richard N. "Soviet Mechanized Airborne Forces." *Infantry* 75 (May–June 1985): 24–28.
"Army Receives Armored Vehicle." *Journal of the Armed Forces* 102 (5 June 1965): 22.
"Army to Put TOW Missiles on M113s by 1978—Soviets Have Them Now." *Armed Forces Journal International* 114 (November 1976): 16.
Avant, Deborah D. *Political Institutions and Military Change: Lessons from Peripheral Wars*. Ithaca, NY: Cornell, 1994.
Ayers, Kenneth L. "Structuring a Combat Maneuver Battalion." *Armor* 85 (May–June 1976): 50–51.
Bacevich, A. J. *The Pentomic Era: The U.S. Army between Korea and Vietnam*. Washington, DC: National Defense University Press, 1986.
Backofen, Joseph E. "Armor Technology Part V: Crew Survivability." *Armor* 93 (January–February 1984): 21–25.
Baggett, John, Vernon Humphrey, and Richard Sinclair II. "Kiss of the Dreadnought." *Infantry* 62 (January–February 1972): 28–30.
Baird, Niven J. "The MICV Is an Offensive Weapons System." *Armor* 86 (July–August 1977): 27–28.
Banks, William F. "Future Combat Vehicles." *National Defense* 60 (1976): 274–75.
Basalla, George. *The Evolution of Technology*. Cambridge: Cambridge University Press, 1988.
Bastable, Marshall J. "From Breechloaders to Monster Guns: Sir William Armstrong and the Invention of Modern Artillery, 1854–1880." *Technology and Culture* 33 (1992): 213–47.
Baxter, William P. "BMD-1: A Formidable Fighting Machine." *Army* 31 (March 1981): 41–43.
Baxter, William P. "Dismounted Soviet Soldier—A Vulnerable Adversary." *Army* 31 (August 1981): 31–32.
Beaumont, Roger. *War, Chaos and History*. Westport, CT: Praeger, 1994.
"The Begleitpanzer 57mm—A New Combat Vehicle Concept for the 90's." *International Defense Review* 11 (1978): 222–24.
Bell, Raymond E. "A New Role For the ACRV." *Armor* 72 (March–April 1963): 49–52.
Bellany, Ian. "The Tank: A Theoretical Exploration of Its Limitations." *RUSI Journal* 119 (December 1974): 61–63.
Bennett, Larry, and Eugene Fox. "Manage the Monster." *Infantry* 55 (September–October 1965): 64–65.

Berens, Robert J. "APC Improvements." *National Defense* 63 (January–February 1979): 20–21.

Bergerud, Eric M. *Red Thunder, Tropic Lightning: The World of a Combat Division in Vietnam*. Boulder, CO: Westview, 1993.

Bergman, John D., and Alexander F. Wejcicki. "The Challenge of Force Modernization." *Armor* 92 (September–October 1983): 30–33.

Bernier, Robert G. "The Combined Arms Maneuver Battalion: Armor and Infantry Build a New Relationship in Fort Hood Experiment." *Armor* 97 (January–February 1988): 14–19.

Besch, Edwin W. "Charging Wagons, Battle Taxis, APC's and IFV's." *Army* 29 (June 1979): 30–35.

Besch, Edwin W. "Infantry Fighting Vehicles: Their Evolution and Significance." *Marine Corps Gazette* 67 (July 1983): 50–60.

"BFV: Integral Part of the Combined Force." *Officer* 58 (September 1982): 16–18.

Birt, M. C. "Armor Protection for Infantrymen." *Armor* 88 (July–August 1979): 55.

Black, Charles. "14 Proverbs Gleaned from a Damp Foxhole." *Infantry* 78 (January–February 1988): 42.

Boatman, John. "A3 Standard Studied for Bradley Fleet." *Jane's Defense Weekly*, 16 January 1993, 8.

Boatman, John. "Keeping Bradley in Business." *Jane's Defense Weekly*, 16 October 1993, 26–31.

Bobinski, Robert S. "Absurd" (letter). *Infantry* 76 (November–December 1986): 5.

Bolger, Daniel P. *Dragons at War: 2-34 Infantry in the Mojave*. Novato, CA: Presidio, 1986.

Bolger, Daniel P. "Maneuver Warfare Reconsidered," in *Maneuver Warfare: An Anthology*, edited by Richard D. Hooker, Jr. Novato, CA: Presidio, 1993, pp. 19–41.

Bolté, Philip L. "A Case of Foot-Dragging: The Mobile Protected Gun System." *Armed Forces Journal International* 120 (July 1983): 74–76.

Bolté, Philip L. "Full Circle: The Armored Cavalry Platoon." *Armor* 103 (September–October 1994): 35–37.

Bolté, Philip L. "M-2/3 Update." *Armor* 89 (July–August 1980): 37–38.

Bolté, Philip L. "Simulation for Armor: The Unit Conduct of Fire Trainer." *National Defense* 67 (September 1983): 16–19.

Bolté, Philip L. "Tank-Automotive News." *National Defense* 71 (September 1987): 24–25.

Bolté, Philip L. "Tank-Automotive News: Looking Ahead." *National Defense* 71 (May–June 1987): 18–19.

Bolté, Philip L. "Tank-Automotive News: The Other Side of the Testing Coin." *National Defense* 72 (November 1988): 10–11.

Bolté, Philip L. "Testing the Bradley." *National Defense* 67 (September 1983): 17–24.

Bolté, Philip L. "Where Were They When the Paper Was Blank? A Response to the Critics of the Bradley Fighting Vehicle." *National Defense* 66 (April 1982): 11–12.

Boly, William. "The Army's $11-Billion Deathtrap." *Reader's Digest* 123 (August 1983): 125–28.

Boly, William. "The $13 Billion Dud." *California*, February 1983: 62–69, 127–28.

Bonsignore, Ezio. "On Recent Developments in Anti-Tank Warfare." *Military Technology* 13 (December 1989): 60–68.

Bonsignore, Ezio, and Ian Bustin. "Towards 'Electronic Armour': Countermeasure

Systems for AFVs." *Military Technology* 17 (December 1993): 10–18.
Bost, Fred H. "Rifle Squad Leader." *Infantry* 51 (November–December 1961): 4–7.
Boudinot, Burton S. "An Approach to the Scout Vehicle Dilemma." *Armor* 79 (1970): 44–45.
Boudinot, Burton S. "Fast Troops." *Armor* 89 (September–October 1980): 30–35.
Boudinot, Burton S. "A Kingdom for a Tank." *Armor* 96 (May–June 1987): 49.
Boudinot, Burton S. "Schnelle Truppen [*sic*] and the ACMV." *Armor* 73 (March–April 1964): 51–54.
Boudinot, Burton S. "We Must Learn from the Past" (letter). *Infantry* 84 (January–February 1994): 3–4.
Bourjaily, Monte, Jr. "Armor Wonders Mainly in Future." *Army Times*, 16 May 1959, 20.
Bowen, Alfred T. "Why Not a Combined Arms Vehicle?" *Armor* 90 (July–August 1981): 71–72.
Bradford, David G. "Light Infantry." *Infantry* 76 (July–August 1986): 22–28.
Bradford, Zeb B. "Mechanized Infantry on the Modern Armored Battlefield." *Infantry* 65 (November–December 1975): 20–29.
Bradley, Clifford D. "Future Close Combat Vehicles." *Armor* 90 (January–February 1981): 36–41.
Bradley, Clifford D. "The Future IFV." *Infantry* 71 (July–August 1981): 21–27.
Bradley, Clifford D. "Historical Military Cycles." *Armor* 92 (January–February 1983): 21–25.
Bradley, Clifford D. "Weapons versus Armor: A New Approach." *Armor* 91 (July–August 1982): 22–25.
"The Bradley: At Last, A Real Test." *Fortune*, 25 February 1991, 50–55.
"Bradley Corner." *Infantry* 84 (November–December 1994): 7.
Brasier, Steve. "Heavy Battalion Trains." *Infantry* 76 (July–August 1986): 15–17.
Bredin, H. E. W., and C. D. C. Logan. "Problems of Infantry Reorganization." *RUSI Journal* 114 (March 1969): 19–25.
Bremer, James H. "A Squad with Two Wings." *Infantry* 49 (April 1959): 61–64.
Bresseau, Timothy L., and Stanley L. Lentz. "Computer Models in Weapons Design." *National Defense* 58 (July–August 1974): 69–71.
Breuhan, David R. "European Reconnaissance Patrols." *Infantry* 79 (March–April 1989): 19–24.
Brewer, J. H. "MBAV: NATO's Best Chance for a Truly Cooperative Vehicle Program?" *International Defense Review* 10 (1983): 780–82.
Brower, Kenneth S. "Armored Fighting Vehicles and Units for the Future." *RUSI Journal* 126 (September 1981): 64–66.
Brown, D. K. *The Future British Surface Fleet*. Annapolis: Naval Institute Press, 1991.
Brown, Michael E. *Flying Blind: The Politics of the U.S. Strategic Bomber Program*. Ithaca, NY: Cornell, 1992.
Brudvig, Dale K. "A Proposed Divisional Organizational Concept." Fort Leavenworth, KS: U.S. Army Command and General Staff College, 1966; reprinted in part in *Selected Readings in the Development of Combat Divisions*, Fort Leavenworth, KS: U.S. Army Command and General Staff College, 1973.
Bruno, Thomas A., and John T. Broome. "A New Day for Armor or the Last Glimmer of Sunset?" *Armor* 99 (September–October 1990): 7–11.
Bryla, Edward A. "Irrational Weapon System Acquisition." *Armor* 97 (January–February

1988): 44–48.
Budul, Vladimir. "Infantry Fighting Vehicles." *Soviet Military Review* 5 (May 1989): 24
Bunce, Peter L. "Appalled" (letter). *Armor* 88 (November–December 1979): 3.
Burba, Edwin H., Jr. "Commandant's Note." *Infantry* 74 (March–April 1986): 2–4.
Burba, Edwin H., Jr. "Commandant's Note." *Infantry* 76 (July–August 1986): 2–3.
Burba, Edwin H., Jr. "Commandant's Note: Distinctions between Brigade and Battalion Command." *Infantry* 77 (May–June 1987): 2–3.
Burke, James T. "Armored Infantry and Recon Unit Organization." *Armor* 65 (July–August 1956): 15–17.
Burneice, Joseph. "BMP-2 Comments" (letter). *Armor* 95 (May–June 1986): 3.
Burneice, Joseph R. "The Operational Manoevre Group: Concept versus Organization." *Military Technology* 10 (October 1986): 66–79.
Burneice, Joseph R., and Paul A. Hoven. "The Future Close Combat Vehicle Family." *Military Technology* 8 (September 1984): 35–43.
Burns, Jerome J. "Lessons on the BIFV." *Infantry* 80 (January–February 1990): 41–42.
Burton, James G. *The Pentagon Wars: Reformers Challenge the Old Guard*. Annapolis: U.S. Naval Institute, 1993.
Bustion, Ian. "Desert WARRIOR." *Military Technology* 15 (July 1991): 53–54.
Butler, Walter G. "U-COFT Effectiveness." *Infantry* 81 (March–April 1991): 15–18.
Butts, Russell K. "Master Gunner Course." *Infantry* 76 (May–June 1986): 36–37.
Cage, Steven D. "Sustaining the Light Infantry Brigade Command Post." *Infantry* 77 (March–April 1987): 33–35.
Caine, Bruce T. "Antitank Tactics." *Infantry* 63 (July–August 1973): 32–37.
Caine, Bruce T. "Dragoons and Hussars: Tomorrow's Maneuver Battalions." *Armor* 83 (November–December 1974): 24–30.
Caine, Bruce T. "Infantry Training in an Armored Division." *Infantry* 63 (March–April 1973): 28–31.
Canby, Stephen L. "Light Infantry in Perspective." *Infantry* 74 (July–August 1984): 28–31.
Carey, Pete. "Rejected Parts Allegedly Used in Bradley." *Mercury News* (San Jose, CA), 5 October 1986.
Cavell, Paul. "Mobile Defense." *Military Review* 52 (February 1972): 77–82.
Cavezza, Carmen J. "Commandant's Note: Bradley Fighting Vehicle Update." *Infantry* 81 (January–February 1991): 1–2.
Cavezza, Carmen J. "Commandant's Note: Tomorrow's Infantry." *Infantry* 80 (July–August): 1–2.
"Celtic Cross IV: The Light-Heavy Mix (1)." *Infantry* 77 (March–April 1987): 27–32.
Chadick, William G. "New Fighting Vehicle." *Armor* 88 (May–June 1979): 2.
Chamberlain, E. W., III. "ITVs in Bradley Units" (letter). *Infantry* 91 (March–April 1991): 3.
Chapman, Anne W. *The Origins and Development of the National Training Center, 1976–1984*. Fort Monroe, VA: Office of the Command Historian, U.S. Army Training and Doctrine Command, 1992.
Chase, Larry. "LAV 8x8, A Better Choice." *Armor* 99 (November–December 1990): 48.
Chaudrue, Robert G. "Requiem for the Infantry." *Infantry* 68 (May–June 1978): 28–31.
Cherry, Jeff F. "MICV: Future Tool of the Trade for Mechanized Infantrymen?" *Infantry* 62 (March–April 1972): 28–30.
Cherry, Jeff F. "Mounted Combat." *Infantry* 65 (September–October 1975): 12–15.

"Chief of Infantry Update." *Infantry* 81 (July–August 1991): 7–8.

Christensen, Dee C. "Dragon Training." *Infantry* 80 (January–February 1991): 37–39.

Clark, Asa A., IV, Peter W. Chiarelli, Jeffrey S. McKitrick, and James W. Reed, eds. *The Defense Reform Debate: Issues and Analysis*. Baltimore, MD: Johns Hopkins, 1984.

Clark, Mark Edmond. "Infantry—Defending Against Armored and Mechanized Attacks." *Military Review* 71 (July 1991): 55–63.

Clarke, Bruce C. "An Estimate of the Armor Situation." *Armor* 95 (November–December 1986): 34–36.

Clarke, Bruce C. "When the Armored Personnel Carrier Was Saved for the Army." *Armor* 80 (September–October 1971): 26.

Clemens, John. "Waking Up from the Dream: The Crisis of Cavalry in the 1930's." *Armor* 99 (May–June 1990): 20–23.

Coates, James, and Michael Kilian. *Heavy Losses: The Dangerous Decline of American Defense*. New York: Viking, 1985.

Coble, Donald W. "Is It Taps for Tanks?" *Armed Forces Management* 16 (February 1970): 48.

Coghlan, Jim. "Bradley Still Drawing Fire." *Defense Electronics* 18 (1986): 86.

Cole, Stephen V. "Firing Port Weapons." *Infantry* 78 (May–June 1988): 5.

Conway, Paul L. "Upgrade M113 Units" (letter). *Infantry* 76 (November–December 1986): 5–6.

Corbett, Richard L. "Loading Plan for M113." *Infantry* 52 (March–April 1962): 27.

Coroalles, Anthony M. "Weapons, Terrain, and Tactics." *Infantry* 79 (July–August 1989): 38–40.

"Costing Logistics Support for the Scout." *Army Logistician* 3 (May–June 1971): 12–13.

Couch, James B. "On Brigade Reconnaissance" (letter). *Armor* 100 (March–April 1991): 2–3.

Crawford, James E. "Reviving Old Concepts" (letter). *Armor* 102 (November–December 1993): 3

Crawley, Vince. "Ghost Troop's Battle at the 73 Easting." *Armor* 100 (May–June 1991): 7–12.

Crevocouer, P. "The U.S. Choice of a Coaxial Tank Machine Gun." *International Defense Review* 91 (October 1976): 770–72.

Critchley, W. Harriet. "Does Doctrine Precede Weaponry?" *International Review* 33 (summer 1978): 524.

Crow, Charles L. *The J-Series Cavalry Squadron and Its Reconnaissance Mission*. Ft. Leavenworth, KS: School of Advanced Military Studies, 1985.

Crowell, Howard G., Jr., and Jared L. Bates. "Heavy-Light Connection: Division." *Infantry* 74 (July–August 1984): 15–18.

Crutchley, Michael J. "The Light Infantry Division: A Status Report." *Military Technology* 10 (October 1986): 56–65.

Cruz, Cesar. "More on Hunter-Killer Technique" (letter). *Armor* 102 (November–December 1993): 3.

Currie, Thomas P. "CFV Not for Scouts" (letter). *Armor* 89 (March–April 1980): 2.

Czeslik, Knut. "*Auftragstakrik*: Thoughts of a German Officer." *Infantry* 81 (January–February 1991): 10–11.

D'Agostino, John F. "The Bradley Master Gunner." *Infantry* 74 (March–April 1984): 9–10.

Deal, Kenneth L. "The Missing Link: Making the HMMWV Scoutworthy." *Armor* 99

(November–December 1990): 36–38.

Delamain, Charles. "How Long Is the Piece of String? The Armored Reconnaissance Vehicle Defined." *Military Technology* 16 (July 1992): 55–57.

Dell, Nat. "The Squad Leader Today: How Well Does He Lead?" *Soldiers* 29 (December 1974): 33–37.

Demchak, Chris C. *Military Organizations, Complex Machines: Modernization in the U.S. Armed Forces*. Ithaca, NY: Cornell, 1991.

De Palo, William A. "Dismounted Night Attack." *Infantry* 75 (September–October 1985): 26–29.

De Palo, William A., Jr. "Extended Cross-Attachment." *Infantry* 74 (July–August 1984): 5–6.

DePuy, William E. "Infantry Combat." *Infantry* 80 (March–April 1990): 8–13.

DePuy, William E. "One-Up and Two-Back?" *Army* 30 (January 1980): 20–25.

DePuy, William E. "The U.S. Army: Are We Ready for the Future?" *Army* (September 1978): 22–29.

"Developing Army's Bradley: The Wages of Compromise." *Army* 37 (July 1987): 71.

Dickey, C. A. "Shotgun Marriage." *Military Review* 36 (June 1956): 48–54.

Dittman, John. "A Day in an Armored Personnel Carrier." *Army Digest* 21 (November 1966): 29–31.

Dmitriyev, I. "Operation and Maintenance of Infantry Fighting Vehicles." *Soviet Military Review* 11 (November 1979): 26–27.

"DOD Lists 'Friendly Fire' Casualities." *Jane's Defense Weekly*, 24 August 1991, 9–13.

Donnelly, Tom. "How the Bradley Became a Target." *Military Forum* 4 (March 1988): 49–52.

Doubler, Michael D. *Closing with the Enemy: How GIs Fought the War in Europe, 1944–1945*. Lawrence: University Press of Kansas, 1994.

Doughty, Robert A. *The Evolution of U.S. Army Tactical Doctrine, 1946–76*. Fort Leavenworth, KS: Combat Studies Institute, 1979.

Dowaing, Wayne A. "Light Infantry Integration in Central Europe." *Military Review* 66 (September 1986): 18–29.

Downing, Wayne A. "Reorganizing." *Infantry* 76 (March–April 1986): 22–26.

Doyle, David K. "Must the MBT Go It Alone?" *Armor* 77 (March–April 1968): 34.

Draude, Thomas V. "Combined Arms Warfare." *Marine Corps Gazette* 73 (April 1989): 59–60.

Drew, Dennis M. "Recasting the Flawed Downsizing Debate." *Parameters* (spring 1993): 39–48.

Dubik, James M., and James J. Montana. "FM 100-5: Conceptual Models and Force Design." *Military Review* 64 (July 1984): 16–26.

Dunaway, Robert L. "A Battle Book for the BFV Platoon Leader." *Infantry* 77 (July–August 1987): 12–13.

Dunaway, Roy S. "The Infantry Fighting Vehicle." *Infantry* 67 (July–August 1977): 46–50.

Dunetz, Bryant R. "Tank Gun or Missile?" *Ordnance* 57 (September–October 1972): 147–50.

Dunn, James A., Jr. "Heavy Force, Light Force: A Heavy Force of the 4th ID (Mech) Maneuvers with a Light Task Force at the NTC." *Armor* 96 (September–October 1987): 10.

Dunn, Walter S., Jr. *Hitler's Nemesis: The Red Army 1930–45*. Westport, CT: Praeger,

1994.
Dunnigan, James F., and Raymond M. Macedonia. *Getting It Right: American Military Reforms after Vietnam to the Persian Gulf and Beyond*. New York: William Morrow, 1993.
Dunning, Forest. "Infantry Weapons." *Infantry* 67 (July–August 1977): 16–18.
Dupuy, R. Ernest, and Trevor N. Dupuy. *The Encyclopedia of Military History from 3500 B.C. to the Present*, 2nd rev. ed. New York: Harper and Row, 1986.
Duran, Terry L. "Heavy Mortar Platoon: Stepchild or Hip-Pocket Artillery?" *Infantry* 77 (May–June 1987): 18–20.
Duran, Terry L. "Shortcomings in New TOE." *Infantry* 76 (March–April 1986): 38–41.
Durisian, Simon. *Vietnam Tracks: Armor in Battle 1945–75*. Novato, CA: Presidio, 1983.
"Echo Company: A Vital Player." *Infantry* 81 (September–October 1991): 13–14.
Edwards, James B. "Mobile Armored Infantry." *Ordnance* 53 (January–February 1969): 399–400.
"The Employment of Non-Mechanised Infantry." *RUSI Journal* 125 (December 1980): 56–69.
English, J. A., J. Addicott, and P. J. Kramers, eds. *The Mechanized Battlefield: A Tactical Analysis*. Washington, DC: Pergamon-Brassey's, 1985.
English, John A. "Light Infantry." *Infantry* 74 (November–December 1984): 19–25.
English, John A. *On Infantry*. New York: Praeger, 1981.
English, John A., and Bruce I. Gudmundsson. *On Infantry*, rev. ed. Westport, CT: Praeger, 1994.
"Enlisted Career Notes: BIFV Infantryman." *Infantry* 78 (September–October 1988): 46.
"Enlisted Career Notes: 11M Infantry." *Infantry* 78 (January–February 1988): 47.
"Enlisted Career Notes: 11M NCO Development." *Infantry* 78 (March–April 1988): 45.
Erb, Hasso. "Heavier Armament for IFVs and Reconnaissance Vehicles?" *International Defense Review* 11 (1978): 1101–5.
Erb, Hasso. "The U.S. Army Combat Support Vehicle—Light Armored Reconnaissance Vehicle or Superjeep?" *International Defense Review* 11 (1978): 917–20.
Erickson, John. "Soviet Combined Arms Operations." *Armor* 89 (May–June 1980): 16–21.
Ernst, Carl F., and David M. White. "Bradley Infantry on the Air Land Battlefield." *Infantry* 76 (May–June 1986): 20–24.
Eshel, D. "Infantry Fighting Vehicles." *Defense Update International* 40 (1983): 14–24.
Eshel, David. "Counterattack in the Sinai: 8 October 1973." *Military Review* 73 (November 1993): 54–67.
Eshel, David. "Modern Inter-arms Concepts: IDF Organization: The Yom Kippur War and After." *Defense Update International* 69 (March 1986): 32–39.
Eshel, David. "Profile of Combined Arms." *National Defense* 73 (September 1988): 48–50.
Esposito, Curtis V. "Armor Operations in Built-Up Areas." *Armor* 91 (July–August 1982): 26–30.
Estes, Kenneth W. "LAV—Quo Vadis?" *Marine Corps Gazette* 65 (December 1981): 18–20.
Evans, David. *Sherman's Horsemen: Union Cavalry Operations in the Atlanta Campaign*. Bloomington and Indianapolis: Indiana University Press, 1996.
Evans, J. B. "Armored Infantry." *Military Review* 24 (March 1945): 38–42.

Everson, John C. "The Armored Infantry Battalion: Organized and Trained as It Will Fight." *Military Review* 61 (September 1981): 54–64.

Ewing, Dale T. "A Turreted M-113 Is Suggested" (letter). *Armor* 90 (November–December 1981): 92–93.

Ewing, Dale T. "Why?" (letter). *Armor* 88 (March–April 1979): 2.

"Exclusive AFJ Interview with: Phillip A. Karber, Vice President and General Manager, National Security Programs, The BDM Corporation." *Armed Forces Journal International* 124 (May 1987): 42–44.

"Exclusive AFJ Interview with: Phillip A. Karber, Vice President and General Manager, National Security Programs, The BDM Corporation." *Armed Forces Journal International* 124 (June 1987): 112–14.

Ezell, Edward C. "The Firing Port Weapon: A New Army Small Arm." *National Defense* 65 (August 1980): 16.

Fallows, James. *National Defense*. New York: Random House, 1981.

Farley, Roy W. "Blackhorse Report II." *Armor* 77 (March–April 1960): 5–11.

Fialka, John. "The Bradley: Bells and Whistles." *Washington Monthly*, April 1982: 22–25.

Fialka, John J. "A New Troop Carrier Is Remarkable—So Are Its History and Cost." *Wall Street Journal*, 17 February 1982, 1, 20.

Fickett, Richard K. "Mortars for Light Platoons." *Infantry* 77 (May–June 1987): 15–18.

Fish, Thomas E. "The Infantry Spectrum: Crossing from Light to Mech." *Infantry* 80 (January–February 1980): 39–41.

Fish, Thomas E. "Think Light." *Infantry* 78 (March–April 1988): 8–10.

Fitzmorris, Lawrence B. "Developing Tomorrow's Combat Vehicles." *Armor* 89 (May–June 1980): 22–25.

Fletcher, Robin. "After BMP-2—A Heavy MICV?" *Jane's Soviet Intelligence Review* 1 (September 1989): 410–12.

Fletcher, Robin. "Crewing and Arming the IFV." *Military Technology* 17 (December 1993): 69–72.

Fletcher, Robin. "Fire Support Vehicles: An Elusive Category." *Military Technology* (March 1994): 8.

Fletcher, Robin. "Two for the Price of One—The Case for the Mini-MICV." *Military Technology* 14 (October 1990): 171–77.

Flynn, Karl S., and Joseph Miller. "Hunter-Killer Operations." *Armor* 102 (July–August 1993): 48–50.

Foley, John E. "Observations on Mechanized Infantry." *Infantry* 76 (July–August 1986): 29–33.

Foley, Thomas C. "Meeting Armor's Challenges: The Cavalry." *Armor* 99 (July–August 1990): 4–5.

Forsyth, Robert W., and John P. Forsyth. "The Cheap Shot: The Lesson of the October War and Its Impact on NATO." *NATO's Fifteen Nations* 19 (August–September 1974): 82–83.

Foss, Christopher F., ed. *Jane's Armour and Artillery, 1994–1995*, 15th ed. Alexandria, VA: Jane's Information Group, 1994.

Foss, John W. "Bradley Organization and Tactics." *Infantry* 75 (July–August 1985): 2–3.

Foss, John W. "Heavy-Light Mix." *Infantry* 75 (September–October 1985): 2–3.

Franz, Wallace P. "Air Mechanization: The Next Generation." *Military Review* 72 (February 1992): 59–66.

Friedman, Norman. *Desert Victory: The War for Kuwait.* Annapolis: Naval Institute Press, 1991.

Friedrich, Robert L. "NET." *Infantry* 75 (September–October 1985): 32–36.

Fuller, J. F. C. *Armored Warfare: An Annotated Edition of "Lectures on F.S.R. III (Operations Between Mechanized Forces)."* Harrisburg, PA: Military Service Publishing Company, 1943.

Funk, David L., and Donald C. Snedeker. "Armored Reconnaissance Scout Vehicle Test." *Armor* 84 (September–October 1975): 37–42.

Funk, Paul E. "Future Thrusts." *Armor* 103 (January–February 1994): 47–50.

"The Future of the Infantry." *Infantry* 72 (September–October 1982): 19–21.

Gallop, David L. "The Driver's Thermal Viewer: A DESERT STORM AAR." *Armor* 103 (May–June 1994): 16–17.

Galvin, John R. "Heavy-Light Forces and the NATO Mission." *Infantry* 74 (July–August 1984): 10–14.

Gameros, Charles W. "Scout HMMWV's and Bradley IFV's: Gulf War Provides a Comparison of Scout Vehicles and MTOE's." *Armor* 100 (September–October 1991): 21–25.

Garland, Albert N. "The MICV: What Is It?" *National Defense* 58 (September–October 1973): 146–48.

Garner, Dixon B. "Armor Symposium 1980: An Overview." *Marine Corps Gazette* 64 (July 1980): 26–31.

Geduldig, Aaron L. "How to Cure the FIST-V Blues." *Field Artillery* (October 1991): 64–66.

Geier, Richard P. "Do We Need a Light Tank?" *Armor* 89 (September–October 1980): 13.

Gellert, Frederick J. "Class III Operations." *Infantry* 80 (May–June 1990): 11–13.

"The German Armored Combat Troops." *Infantry* 73 (May–June 1983): 24–27.

Gibbons, Edward G., Jr. "Security Force Operations in a Mechanzed Infantry Task Force." *Infantry* 81 (July–August 1991): 22–26.

Gibbons, Edward G. Jr. *Why Johnny Can't Dismount: The Decline of America's Mechanized Infantry Force.* Ft. Leavenworth, KS: U.S. Army Command and General Staff College, 1995.

Gibson, A. T. "Infantry and Armour, and the Influence of Mobility." *Army Quarterly* 103 (January 1973): 193–99.

Gibson, James M. "A Case for Mechanized Infantry." *Military Review* 50 (September 1970): 56–70.

Gillie, Mildred Hanson. *Forging the Thunderbolt: A History of the Development of the Armored Force.* Harrisburg, PA: Military Service Publishing Co., 1947.

Gilman, Robert L., and Howard W. Crawford, Jr. "Mechanized Infantry Doctrine." *Infantry* 67 (July–August 1977): 39–46.

Glenn, Edmund S. "New Concepts for Armor." *Ordnance* 52 (June–February 1968): 382–85.

Gourley, Scott R. "The BTR: Ivan's Other Carrier." *Infantry* 74 (September–October 1984): 11–13.

Gourley, Scott R. "FMC's Fighting Vehicle System Delivers Army Firepower, Mobility, and Survivability." *Defense Systems Review* 2 (October 1984): 18–21.

Gourley, Scott R. "Lessons For Bradley." *Jane's Defense Weekly* 16 (7 December 1991): 1113.

Gourley, Scott R. "A Workhorse Gets New Power." *Army* 44 (July 1994) 40–44.
Graham, Barry. "Mechanized Forces in Vietman." *Infantry* 59 (July–August 1969): 45–48.
Granger, Clinton E. "Wanted: An Infantry Fighting Vehicle." *Military Review* 43 (February 1963): 26–35.
Gravenstede, Nelson F. "Future Armor Rearm System." *Armor* 100 (July–August 1991): 17–18.
Gray, Colin S. "Does Theory Lead Technology?" *International Journal* 33 (summer 1978): 506–23.
Green, Constance McLaughlin, Harry C. Thomson, and Peter C. Roots. *The Ordnance Department: Planning Munitions for War*. Washington, DC: USGPO, 1955.
Green, Michael. "Better Bradley." *International Combat Arms* 7 (January 1989): 55–59.
Green, Michael R. "The Future Bradley Vehicles." *Armada International* 9 (October 1985): 84.
Greenfield, Kent Roberts, Robert R. Palmer, and Bell I. Wiley. *The Organization of Ground Combat Troops*. Washington, DC: USGPO, 1947.
Gregory, William H. *The Price of Peace: The Future of the Defense Industry and High Technology in the Post–Cold War World*. New York: Maxwell Macmillan, 1992.
Greve, Frank. "Dream Weapon a Nightmare." *Inquirer* (Philadelphia, PA), 2 May 1982; reprinted in *Parameters* 12 (September 1982): 76–82.
Griest, R. "FVS—A True Mounted Infantry Combat System." *International Defense Review* 11 (1978): 1, 79–80.
Griffith, Craig A. "Let's Get the Right Vehicles." *Marine Corps Gazette* 65 (March 1981): 20–21.
Griffith, Paddy. *Battle Tactics of the Western Front: The British Army's Art of Attack, 1916–1918*. New Haven, CT: Yale University Press, 1994.
Gritz, John P. "Light Armored Vehicles or Light Armored Victims?" *Marine Corps Gazette* 66 (August 1982): 36–42.
Gutzman, Phillip C. "The Bradley Saves Ivan." *National Guard* 39 (August 1985): 32–36.
Hackney, Michael S. "Echo Company: The Fifth Player." *Infantry* 75 (July–August 1985): 20–24.
Hackney, Michael S. "Understrength Platoon." *Infantry* 72 (November–December 1982): 10–11.
Hades, John T. "Carrier Battle Drill." *Infantry* 48 (April 1958): 28–33.
Halbert, Gerald A. "Bradley against the Opposition." *Defense Update International* 41 (1983): 3–18
Halbert, Gerald A. "The Bradly [*sic*] versus the Opposition." *Armor* 91 (May–June 1982): 26–29.
Hale, Stephen Delaney, and John Winters. "Aiken Tank Plant Boasted; Cuts Stand." *Chronicle* (Augusta, GA), 7 October 1992.
Hall, Robert. "Possible Future Developments in the Combined Arms Structure of the Soviet Ground Forces." *Jane's Soviet Intelligence Review* 1 (March 1978): 123–25.
Hall, Wayne M. "The Active Defence." *Armor* 89 (May–June 1980): 12–15.
Hallenbeck, Ralph A. "Reorganize Platoon." *Infantry* 73 (November–December 1983): 10–12.
Hallion, Richard P. *Girding for War: Perspectives on Research, Development, Acquisition, and the Decision-Making Environment of the 1980's*. Edwards AFB, CA:

History Office, Air Force Flight Test Center, 1985.
Hallion, Richard P. *The Historical Evolution of Commonality in Fighter and Attack Aircraft Airframe Development and Usage*. Edwards AFB, CA: History Office, Air Force Flight Test Center, 1985.
Halloran, Barney. "Battle Taxi or MICV?" *Soldiers* 29 (January 1974): 28–33.
Hamilton, Michael A. "Heavy-Light Operations." *Infantry* 79 (September–October 1989): 22–26.
Harris, J. P. *Men, Ideas and Tanks: British Military Thought and Armored Forces. 1903–1939*. Manchester: Manchester University Press, 1995.
Harris, J. P., and F. H. Toase, eds. *Armoured Warfare*. New York: St. Martin's Press, 1990.
Hart, Gary, and William S. Lind. *America Can Win: The Case for Military Reform*. Washington, DC: Adler and Adler, 1986.
Hartzog, William W., and John D. Howard. "Heavy/Light Operations." *Military Review* 67 (April 1987): 24–33.
Harvey, Andrew, and Robert Firkins. "The Mk 19 Mod3 Grenade Machine Gun on the M1 and M3." *Armor* 100 (September–October 1991): 34–35.
Haszard, Sidney S. "The Gut Issue." *Armor* 87 (May–June 1978): 50.
Havrell, Ben. "A New Organizational Concept." *Infantry* 53 (January–February 1963): 2–7.
Hayes, Joseph B. "Change in the MICV Program." *National Defense* 56 (January–February 1977): 325–26.
Hayes, Joseph B. "MICV Plus TOW." *National Defense* 58 (March–April 1975): 358.
Hayes, Joseph B. "MICV Program." *Ordnance* 52 (May–June 1973): 442–44.
Hayes, Joseph B. "MICV Requests Released at Last." *Ordnance* 52 (September–October 1972): 108.
Hayes, Joseph B. "MICV Survives." *National Defense* 59 (January–February 1975): 336.
Hayes, Joseph B. "M113 Modernization Program." *National Defense* 57 (July–August 1977): 16–17.
Hayes, Joseph B. "Scout Bites the Dust." *Armed Forces Journal International* 112 (October 1974): 29–30.
Hayes, Joseph B. "Scout Bites the Dust." *National Defense* 53 (July–August 1974): 26f.
Hayes, Joseph B. "Selection for Scout." *Ordnance* 51 (May–June 1972): 450–52.
Hayes, Joseph B. "To Develop MICV." *Ordnance* 52 (January–February 1973): 276–77.
Hayes, Joseph B. "Tracks or Wheels?" *Ordnance* 51 (January–February 1972): 274.
Hayes, Joseph B. "The U.S. Army Scout Vehicle Competition." *International Defense Review* 6 (February 1973): 86–90.
Hayes, Joseph B. "User/Developer Conference." *Ordnance* 52 (July–August 1972): 22–24.
Hayes, Joseph B. "Will Scout Hurt MICV?" *Ordnance* 51 (March–April 1972): 408.
Hayes, W. H. "M3 as TOW Vehicle" (letter). *Infantry* 81 (July–August 1991): 5–6.
Hayes, William H. "Revamp Echo Company" (letter). *Infantry* 76 (January–February 1986): 6–7.
Helena, Marshall L. "Light to Mech" (letter). *Infantry* 80 (May–June 1990): 3.
Hemphill, John A. "Reorganization of the Mechanized Infantry Battalion." *Infantry* 65 (September–October 1975): 18–27.
Henderson, William Darryl. *Cohesion: The Human Element in Combat*. Washington, DC: National Defense University Press, 1985.

Henderson, William Darryl. *The Hollow Army: How the U.S. Army Is Oversold and Undermanned*. Westport, CT: Greenwood, 1990.

Henry, Peter Albert. "Rebuttal on 'A New Concept'." *Armor* 93 (September–October 1984): 2–3.

Herbert, Paul H. *Deciding What Needs to Be Done: General William DePuy and the 1976 Edition of FM 100-5, Operations*. Fort Leavenworth, KS: Combat Studies Institute, U.S. Army Command and General Staff College, 1988.

Herrly, Peter F. "The Army's Light Divisions: Where Next?" *Military Review* (January 1994): 78–60.

Herrly, Peter F. "Middleweight Forces and the Army's Deployability Dilemma." *Parameters* 18 (September 1989): 46–59.

Hewitt, William D. "Rethinking the Movement to Contact." *Armor* 93 (January–February 1984): 15–20.

HHC 1/8 Cavalry Scout Platoon. "Scouts Need the New Bradleys" (letter). *Armor* 100 (March–April 1991): 3–4.

Hidula, Scott. "Live Fire Test Results Expected Soon for FMC's Bradley." *Business Journal* (San Jose, CA), 15 June 1987.

Hillen, John. "2d Armored Cavalry: The Campaign to Liberate Kuwait." *Armor* 100 (July–August 1991): 8–12.

Holley, I. B., Jr. *Ideas and Weapons: Exploitation of the Aerial Weapon by the United States during World War I*. New Haven, CT: Yale University Press, 1953.

Holliday, David C. "The M60 AX: A Solution or a Problem?" *National Defense* 60 (March–April 1981): 10–17.

Holliday, David C. "Tank-Automotive News." *National Defense* 60 (May–June 1981): 189–90.

Hollingsworth, James. "Remembering Some Hard-Fought Lessons of WWII." *Armor* 98 (January–February 1989): 6.

Hollingsworth, James. "Remembering Some Hard-Fought Lessons of WWII (Part II)". *Armor* 98 (March–April 1989): 5–6, 45–46.

Hollingsworth, James F., and Allan T. Wood. "The Light Armored Corps—A Strategic Necessity." *Armed Forces Journal International* 117 (January 1980): 20–22.

Holmes, David G. "The IFV on the Modern Battlefield." *Infantry* 70 (September–October 1980): 10–14.

Holmes, David G. "Light Heavyweight." *Armor* 88 (May–June 1979): 35–38.

Holmes, David G. "The U.S. Army's Infantry and Cavalry Fighting Vehicles." *International Defense Review* 13 (1980): 1075–81.

Hooker, Richard D., Jr., ed. *Maneuver Warfare: An Anthology*. Novato, CA: Presidio, 1993.

Hooker, Richard D. "The Mythology Surrounding Maneuver Warfare." *Parameters* 12 (spring 1983): 27–38.

Horncastle, Court R. "Scouts Must Train Reconnaissance [*sic*] by Fire" (letter). *Armor* 100 (July–August 1991): 49.

House, Jonathan M. "Designing the Light Division, 1935–1944." *Military Review* 62 (May 1984): 39–47.

House, Jonathan M. *Toward Combined Arms Warfare: A Survey of 20th Century Tactics, Doctrine, and Organization*. Fort Leavenworth, KS: U.S. Army Command and General Staff College, 1984.

Hovell, Bergen B. "Armored Infantry Is Different." *Army* 9 (November 1958): 49–53.

Hoven, Paul A., and Joseph R. Burniece. "IFV-BMP Comparison." *Armor* 91 (September–October 1982): 4.

Howard, Bart. "The HMMWV Scout Platoon" (letter). *Armor* 98 (July–August 1989): 3.

"*Human Factors in Mechanized Warfare*, by Richard E. Simpkin" [book review]. *Infantry* 74 (March–April 1984): 46.

Humphrey, Vernon W. "The Dragoon Concept." *Military Review* 52 (January 1972): 17–25.

Humphrey, Vernon W. "Mech Infantry—What Next?" *Infantry* 61 (July–August 1971): 46–49.

Hunnicutt R. P. *Abrams: A History of the American Main Battle Tank, Volume II.* Novato, CA: Presidio, 1990.

Hunnicutt, R. P. *Sherman: A History of the American Medium Tank.* Belmont, CA: Taurus, 1978.

Hunnicutt, R. P. *Stuart: A History of the American Light Tank, Volume I.* Novato, CA: Presidio, 1992.

Hunt, Timothy P. "Infantry and Armor Cross-Attachments Enhance Combat Readiness." *Armor* 96 (March–April 1987): 43–44.

Huppert, G. Harry. "Mechanized Infantry." *Infantry* 51 (September–October 1961): 62–63.

Huppert, G. Harry. "The 1966 Squad?" *Infantry* 51 (January–February 1961): 35–37.

Hutton, W. M. "The Future of Armour." *RUSI Journal* 113 (May 1968): 95–103.

Hyatt, John Alan. "The LAV-25 in the Scout Role." *Armor* 100 (September–October 1991): 32–33.

Icks, Robert J. "Questions Concerning the CFV" (letter). *Armor* 90 (July–August 1981): 3.

"Infantry Division (Light)." *Infantry* 74 (March–April 1989): 14–16.

"Infantry News." *Infantry* 73 (January–February 1983): 3–5.

"Infantry News." *Infantry* 75 (September–October 1985): 8.

"Infantry News." *Infantry* 76 (May–June 1986): 8.

"Infantry News." *Infantry* 77 (November–December 1987): 7.

"Infantry News." *Infantry* 78 (March–April 1988): 5.

"Infantry News." *Infantry* 80 (March–April 1990): 5.

"Infantry News." *Infantry* 80 (July–August 1990): 7.

"Infantry News." *Infantry* 81 (January–February 1991): 6–8.

"Infantry News: Chief of Infantry Update." *Infantry* 81 (September–October 1981): 6–7.

Inge, Joseph R., and James P. Isenhower, Jr. "Combined Arms: A Look Ahead." *Armor* 85 (July–August 1976): 9–12.

Isby, David C. *Weapons and Tactics of the Soviet Army.* London: Jane's, 1981.

Jacobson, Michael R. "Antiarmor: What You Don't Know Could Kill You." *Infantry* 80 (March–April 1990): 37–40.

Jacobson, Michael R. "Iraqi Infantry." *Infantry* 81 (January–February 1991): 33–37.

Jacobson, Michael R. "Soviet ATGM Countermeasures." *Infantry* 81 (March–April 1992): 8–10.

Janicki, Edward. "Future Military Vehicles." *Ordnance* 52 (May–June 1968): 578–81.

Jarrett, Joseph G. "Combined Arms—Lessons Learned." *Armor* 93 (November–December 1984): 42–43.

Jeffries, Lewis I. "A Blueprint for Force Design." *Military Review* 71 (August 1991): 20–31.

Jenkins, D. H. C. "Autoloading in Tanks." *International Defense Review* 17 (1984): 907–18.

Jenkins, D. H. C. "The MOWAG Tornado MICV." *International Defense Review* 4 (1981): 469–71.

Jenkins, D. H. C. "Operating the Bradley Infantry Fighting Vehicle." *International Defense Review* 15 (1982): 1203–6.

Jennings, Joe. "The Combat Vehicle Technology Program." *Armor* 90 (July–August 1981): 45–46.

Jett, John A. "Light Scouts: Reinventing an Infantry Idea?" *Armor* 98 (September–October 1989): 2–3.

Johnston, John C. "The Journey to Force XXI's Mounted Component." *Armor* 103 (March–April 1994): 14–16.

Johnston, Richard G. "Flaws Seen in Light Scout Concept" (letter). *Armor* 98 (September–October 1989): 3.

Kafkalas, Peter N. "Light Divisions and Low-Intensity Conflict: Are They Losing Sight of Each Other?" *Military Review* 66 (January 1986): 19–27.

Karbeling, Emanuel. "MICV Update." *Infantry* 64 (March–April 1974): 7–9.

Karber, Phillip A. "Anti-Tank Weapons and the Future of Armor: Soviet Thought and NATO's Procrastination." *Armed Forces Journal International* 114 (November 1976): 20.

Karber, Phillip A. "The Growing Armor / Anti-Armor Imbalance in Central Europe." *Armed Forces Journal International* 118 (July 1981): 37–40.

Karber, Phillip A. "The Soviet Antitank Debate." *Armor* 85 (November–December 1976): 10–14.

Karber, Phillip A. "The Soviet Antitank Debate." *Survival* 18 (May–June 1976): 105–11.

Karemaa, Aada. "The Airmobile Armored Division." *Military Review* 50 (September 1970): 3–8.

Keenan, John M. "Developing Tomorrow's Combat Vehicles—Part II." *Armor* 90 (March–April 1981): 42–44.

Kehoe, Thomas P. "The M113 Armored Porcupine." *Infantry* 56 (January–February 1966): 22–26.

Kelly, Henry E. "Rifleman or LWI?" *Infantry* 53 (November–December 1963): 30–32.

Kelly, Henry E., and Frank L. Brown. "The Quick or Dead." *Infantry* 53 (March–April 1963): 50–51.

Kennedy, Donald R. "Ballistics Report." *Armor* 95 (July–August 1986): 4–5.

Kennedy, Donald R. "Improving Combat Crew Survivability." *Armor* 92 (July–August 1983): 16–22.

Kennedy, Edwin L. "Israeli M113s." *Infantry* 74 (July–August 1984): 6–7.

Kennedy, William V. "The New Armored Warfare." *Ordnance* 47 (July–August 1962): 48–50.

Killabrew, Robert B. "Has Light Infantry Really Had Its Day?" *Army* 29 (December 1974): 44–48.

Kindsvatter, Peter S. *The Army of Excellence Divisional Cavalry Squadron—A Doctrinal Step Backward?* Ft. Leavenworth, KS: School of Advanced Military Studies, U.S. Army Command and General Staff College, 1985.

King, D. E. "The Survival of Tanks in Battle." *RUSI Journal* 123 (March 1978): 26–31.

King, Marc A. "Alive and Well." *Armor* 88 (September–October 1978): 35–37.

King, Marc A. "Suited to Scout." *Armor* 89 (March–April 1980): 47–48.

King, Marc A. "Time to Get on Board." *Armor* 89 (March–April 1980): 3.

King, Marc A. "2 x 2: The Regimental Cavalry Troop." *Armor* 90 (March–April 1981): 12–13.

Kirtland, Daniel L. "Some Cavalry MTO&E Changes." *Armor* 95 (May–June 1986): 3.

Kitfield, James. *Prodigal Soldiers: How the Generation of Officers Born of Vietnam Revolutionized the American Style of War*. New York: Simon and Schuster, 1995.

Klaffus, Ernst. "The Infantry—Thoughts on Weapons and Protection." *NATO's Sixteen Nations* 28 (December 1983–January 1984): 36.

Klein, William E. "Mechanized Infantry in Vietnam." *Infantry* 61 (March–April 1971): 18–21.

Knapp, George E. "Echo on the Battlefield." *Infantry* 25 (September–October 1985): 30–33.

Koburger, Charles. W. "Armored Infantry Is Different." *Armor* 60 (January–February 1951): 40–41.

Koch, Gero. "German Infantry in the 1990's." *Infantry* (July–August 1987): 27–32.

Kojro, Chester A. "Bradley Platoon Organization." *Infantry* 77 (March–April 1987): 16–18.

Kolacheski, Richard F. "Division Restructuring." *Armor* 87 (November–December 1978): 18–23.

Kopacz, Joseph C. "The Cadre Armor Division Concept." *Armor* 100 (November–December 1991): 41–47.

Krapke, Paul-Werner. "What Is the Ideal Armoured Personnel Carrier? How Best to Combine Armour with Mechanized Infantry." *Armada International* 12 (June 1988): 21–22.

Krawciw, Nicholas S. H. "A Challenge for Armor." *Armor* 73 (January–February 1964): 52–54.

Lancaster, Michael S. "The Armor Force in the AirLand Battle." *Armor* 91 (January–February 1982): 26–32.

Lartique, Louis J. "Fixing the Scout Platoon" (letter). *Armor* (January–February 1995): 51.

Lawrence, James C. "The Plight of the Scout Vehicle" (letter). *Armor* (March–April 1994): 4.

Leahy, James B., Jr. "Killing Enemy Armor." *Infantry* 79 (November–December 1989): 8–11.

Lehmann, Douglas K., and Robert J. O'Neil. "Two Views on: The Rifle Squad." *Infantry* 70 (May–June 1980): 18–21.

Lenihan, James. "Logical Antiarmor Training." *Infantry* 74 (March–April 1985): 40–41.

Leonard, R. W. "The 'Queen' Makes Tracks." *Armor* 63 (September–October 1954): 6–12.

Leone, Lawrence A. "The Bradley: Let's Make It Even Better." *Infantry* 81 (November–December 1991): 13–14.

Leonhard, Robert R. *The Art of Maneuver: Maneuver-Warfare Theory and AirLand Battle*. Novato, CA: Presidio, 1991.

Leslie, John N. "What Infantrymen and Tankers Need to Know about Serving as Armored Cavalrymen." *Armor* 96 (January–February 1987): 18–21.

Lessels, Bob. "Second-Generation Bradley's [*sic*] Arrive at APG for Testing." *AMC News* (November 1984): 10.

Leuer, Kenneth C. "Commandant's Note: Standardizing Our Units." *Infantry* 78

(September–October 1988): 1.

Leuer, Kenneth C. "Infantry Doctrine—What Should It Be?" *Infantry* 77 (September–October 1987): 1–2.

Liddell Hart, B. H. *The Tanks*. London: Cassell, 1959.

Liddell Hart, B. H. "Will the Tank Survive?" *Ordnance* 44 (May–June 1960): 870–72.

Lind, William S. "Defining Maneuver Warfare for the Marine Corps." *Marine Corps Gazette* 64 (March 1980): 55–58.

Lind, William S. "FM 100-5, Operations: Some Doctrinal Questions for the U.S. Army." *Military Review* 57 (March 1977), 54–65.

Lind, William S. "Mission and Force Structure." *Marine Corps Gazette* 59 (December 1975): 11–16.

Lind, William S. "Proposing Some New Models for Marine Mechanized Units." *Marine Corps Gazette* 62 (September 1978): 34–38.

Lindsay, James J. "The Infantry Division (Light)." *Infantry* 74 (January–February 1984): 2–3.

Livsey, William J. "The Active Defense." *Armor* 89 (March–April 1980): 49–50.

Lockhart, Christopher G. "Modern Dragoons: Bradley Mechanized Infantry." *Infantry* 82 (November–December 1992): 33–35.

London, Herbert I. *Military Doctrine and the American Character: Reflections on AirLand Battle*. New York: National Strategy Information Center, 1985.

Lorber, Azriel K. "Armored Infantry: The Real One." *Armor* 95 (November–December 1986): 40–42.

Lucas, James, and Matthew Cooper. *Panzer Grenadiers*. London: Macdonald and Jane's, 1977.

Ludvigsen, Eric C. " 'Get-Well' Point—Still Over Horizon." *Army* 31 (October 1981): 240–49.

Ludvigsen, Eric C. "IFV." *Army* 29 (June 1979): 26–30.

Ludvigsen, Eric C. "Mechanized Forces in the Next Century." *Army* 37 (July 1987): 22–26.

Lunt, J. D. "Too Few Infantry?" *RUSI Journal* 104 (February 1959): 88–92.

Luttwak, Edward. *The Pentagon and the Art of War*. New York: Institute for Contemporary Studies / Simon and Schuster, 1984.

Lvov, Sergei. "IFV's in the Mountains." *Soviet Military Review* 9 (September 1987): 45.

Lyach, G. G. "Give Infantry More Fire Power." *Combat Forces* 2 (August 1951): 17–18.

Lyle, James, ed. "Division 86." *Army* 33 (July 1983): 18–22; *Army* 33 (August 1983): 18–22; *Army* 33 (September 1983): 26–31; *Army* 33 (November 1983): 34–37; *Army* 33 (December 1983): 34–38; *Army* 34 (February 1984): 14–26.

McCaffrey, Barry R. "Infantry on the High–Intensity Battlefield of Central Europe." *Military Review* 58 (January 1978): 3–6.

MacCaskill, Douglas C. "The Marine Rifle Squad." *Marine Corps Gazette* 55 (April 1971): 40–43.

McCaslin, James K. "The BMP in Combat." *Infantry* 67 (January–February 1977): 26–31.

McClusky, James F. "Infantry Combat Vehicles." *National Defense* 58 (May–June 1974): 536–37.

McColl, Alexander M. S. "Commando Square—The Infantry Battalion Revisted." *Military Review* 55 (April 1975): 52–55.

McCurry, William K., and Phillips, Joel R. "Brigade Deep Operations: Task Organizing for Victory." *Armor* 83 (September–October 1993): 42–45.

McDermott, David F., and Scott R. Gourley. "Evolution of the BMP." *Armor* 73 (November–December 1983): 19.

Mace, James E. "The Combined Arms Battalion: Reality or Myth?" *Armor* 83 (July–August 1974): 22–26.

McGrath, John J. "Combined Arms: An Author's Reply." *Armor* 94 (March–April 1985): 3–4.

McGrath, John J. "A New Concept for Combined Arms." *Armor* 93 (January–February 1984): 46–47.

Mack, Richard E. "The Light Infantry Is too Heavy." *Infantry* 62 (May–June 1972): 14–15.

Mackey, Bruce D. "Doctrinal Publications." *Infantry* 75 (March–April 1985): 38–40.

McLaughlin, Mark W. "Airmobile Operations for Mechanized Infantry Units." *Infantry* 79 (July–August 1989): 40–42.

McMichael, Scott R. *A Historical Perspective on Light Infantry*. Fort Leavenworth, KS: U.S. Army Command and General Staff College, 1987.

McMichael, Scott R. "Three Kinds of Infantry" (letter). *Infantry* 75 (November–December 1985): 4.

McNaugher, Thomas L. *New Weapons, Old Politics: America's Military Procurement Muddle*. Washington, DC: Brookings, 1989.

McVey, Peter M. *Sending the Very Best: An Oral History Interview with Major General Peter M. McVey, Program Executive Officer, Armored Systems Modernization*. Warren, MI: Historical Office, U.S. Army Tank Automotive Command, 1993.

Maginnis, Robert L. "ITV and IFV Transition." *Infantry* 71 (November–December 1981): 18–21.

Malcolm, Ben S. "Infantry: Rolling Against the Punch." *Infantry* 62 (May–June 1972): 6–9.

Malone, Daniel K. "The Infantry Fighter Vehicle." *National Defense* 64 (July–August 1979): 40–45.

Malone, Paul B., IV. "Regiments to the Rescue." *Military Review* 72 (August 1994): 78–80.

Mans, R. S. N. "Light Armored Vehicles—The Way Ahead." *RUSI Journal* 127 (September 1982): 58–62.

Mans, R. S. N. "Whither the Tank?" *Army Quarterly* 104 (April 1974): 290–94.

Mansoor, Peter R. "Basing Light Armor Battalions." *Armor* (November–December 1993): 15–16.

"Marder—West Germany's New APC Enters Service." *International Defense Review* (August 1971): 366–71.

Marriott, John. "The Anti-tank Problem." *NATO's Fifteen Nations* 17 (April–May 1972): 72–82.

Marriott, John. "Fighting Vehicles." *NATO's Fifteen Nations* 21 (April–May 1976): 59–60.

Marzloff, Jean. "The AMX 10P Infantry Combat Vehicle." *International Defense Review* (August 1971): 372–75.

Matheny, Michael R. "A Cure for the 'Cross-Attachment Blues.' " *Armor* 90 (November–December 1981): 51–52.

Maynard, Wayne K. "Spears vs. Rifles: The New Equation of Military Power." *Parameters* 19 (spring 1993): 49–58.

Mazarr, Michael J. *Light Forces and the Future of U.S. Military Strategy*. New York: Brassey's, 1990.

Mazarr, Michael J. "Light-Heavy Debate Rears Its Head Again." *Armed Forces Journal International* 127 (October 1989): 99.

Mazarr, Michael J. "Middleweight Forces for Contingency Operations." *Military Review* 71 (August 1991): 32–39.

Meaders, Mark A., and Rick Baillergeon. "The Maintenance Battle." *Infantry* 80 (May–June 1990): 14–16.

Meadows, Charles L. "Defining the Infantry Mission." *Marine Corps Gazette* 58 (June 1974): 50–51.

Meller, Rudi. "The Fire Support Combat Vehicle: A Powerful New Member of the M113 Family." *International Defense Review* 12 (1979): 252–55.

Merriman, Robert C. "On Cohesion, as in Combined Arms" (letter). *Armor* 97 (July–August 1988): 2–3.

Merriman, Robert C., Jr. "Team Vehicles" (letter). *Infantry* 77 (May–June 1987): 6.

Merzlyak, L. "IFV's on the Offensive." *Soviet Military Review* 7 (July 1987): 18–19.

Mesenger, Charles. *The Art of Blitzkrieg*. London: Ian Allan, 1991.

Meyer, E. C. *E. C. Meyer, General, U.S. Army Chief of Staff, June 1979–June 1983* [s.l.; s.n.; 1983 or 1984].

Meyer, Harold J. "The Infantry Needs Mobility—Not Armor." *Infantry* 49 (April 1959): 65–66.

Meyer, Harold J. "Weight—Nemesis of Mobility." *Infantry* 49 (July 1959): 8–12.

Meyer, Ronald J. "German Armor Today." *Armor* 75 (September–October 1966): 34–38.

"MICV Takes New Direction, Now TBAT-II." *Armed Forces Journal International* 114 (May 1977): 11.

Miksche, Ferdinand Otto. "Effective Antitank Defense." *Military Review* 51 (December 1971): 48–56.

Miles, Donna. "Benning's Bradley U." *Soldiers* 44 (April 1989): 28–31.

Miller, D. M. O. "The Infantry Combat Vehicle: An Assessment." *Military Technology* 9 (May–June 1979): 29–38.

Miller, Daniel B. "More Thoughts on Light Armor" (letter). *Armor* 96 (July–August 1987): 4–5.

Miller, Don. "The Handlebar Infantry." *Army* 30 (September 1980): 38–40.

Miller, John W., III. "Cleaning the Defile: A Doctrinal Discussion." *Armor* (November–December 1994): 26–32.

Miller, Joseph K. "The Platoon Team." *Infantry* 76 (January–February 1986): 14–16.

Miller, Kenneth A. "Making the Bradley Swim." *Army Logistician* (May–June 1988): 18–21.

Miller, Martin J. "Soviet Personnel Carriers." *Ordnance* 53 (January–February 1969): 394–97.

Miller, Martin J. ,Jr. "Russian Combat Vehicles." *Ordnance* 51 (July–August 1966): 49–54.

Miller, Stephen W. "New Trends in Armoured Fighting Vehicles." *Military Technology* 15 (July 1991): 10–17.

Millett, Allan R. "Military Reform in America." *Air University Review* 36 (September–October 1985): 2–10.

Millis, Walter. "Trojan Horse on Tracks." *Armor* 63 (March–April 1954): 28–29.

Milstead, Cole. "Future Heavy Forces: The Need for Better Air Deployability." *Armor*

100 (May–June 1991): 38.

Mintz, Howard. "Ex-Employee Says FMC Hid Bradley Faults." *Peninsula Times Tribune* (Palo Alto, CA), 18 September 1986.

Mitchell C. C. "A Future for the Infantry Arm." *RUSI Journal* 111 (August 1966): 225–29.

Mokyr, Joel. *The Lever of Riches*. Oxford: Oxford University Press, 1990.

Monteverde, Roberto, and Maurilio Tamaio. "The Soviet Combined Arms Reinforced Battalion." *Military Technology* 9 (October 1985): 180–87.

Moore, Ronald W. "High-Firing Bradley." *Soldiers* 42 (February 1987): 32.

Morison, Elting E. *Men, Machines, and Modern Times*. Cambridge MA: MIT Press, 1966.

Morley, Joe. "AOE Not Working" (letter). *Infantry* 81 (March–April 1991): 2–3.

Morley, Thomas V., and Anthony J. Tata. "The Mechanized Infantry Team in the Offense." *Infantry* 80 (May–June 1990): 16–19.

Morrow, R. C. "Does the Fire Team Need Reorganizing?" *Marine Corps Gazette* 53 (February 1969): 28–30.

Mosher, Craig C. "The CFV as a Scout Vehicle Is Questioned" (letter). *Armor* 91 (January–February 1982): 3.

Moskalyou, Vladimir, and Vladimir Taranenko. "APC's in the Mountains." *Soviet Military Review* 5 (1986): 30–31.

Moten, Matthew. "CAMBs: A Better Solution." *Infantry* 78 (September–October 1988): 13–15.

Motley, James B. "Heavy-Light Forces: Assessing the Challenge." *Infantry* 75 (1985): 13–14.

Mouze, Victoria. "The Infantry Moves into the 21st Century." *Soldiers* 39 (June 1984): 44–46.

"M2A2/M3A2 Bradley Fighting Vehicles." *Army* 41 (June 1991): 52–54.

Mundstock, Jack E. "Just One." *Infantry* 76 (March–April 1986): 7.

Naisawald, L. Van L. "The U.S. Infantry Division: Changing Concepts in Organization, 1900–1937," in *Development of Combat Divisions*. Fort Leavenworth KS: U.S. Army Command and General Staff College, 1971.

Nall, R. W. "Combined Arms Operations—Ours Versus Theirs." *Armor* 89 (November–December 1980): 50–51.

Name Withheld. "Do Right by the Bradley." *Army Times*, 5 December 1994, 34.

Nash, Trevor. "The Dynamic Defender: CV90." *Military Technology* 15 (July 1991): 83–88.

Naylor, Sean D. "ATACMS, Comanche Star in Exercise." *Army Times*, 28 November 1994, 22.

Negin, Jonathan J. "DESERT STORM—The First Firefight." *Armor* 94 (March–April 1994): 8–9.

"New Army Squad May Mean $4.6 Billion in New Gear Is Designed to Wrong Size." *Armed Forces Journal International* 114 (March 1977): 15.

"New Scout Platoon Concept Will Test HMMWVs as 'Stealthy' Scouts." *Armor* 98 (March–April 1989): 51.

"New Soviet Armor" *Armor* 77 (July–August 1968): 32–33.

"New Turret, 35mm Cannon for Bradleys." *Armor* 94 (July–August 1985): 49.

Newell, Clayton R. *The Framework of Operational Warfare*. London: Routledge, 1991.

Newell, Clayton R. "Heavy Light Forces: Divisions or Brigades?" *Infantry* 75 (January–

February 1985): 12–13.

Newman, A. S. "Infantry in Retrospect." *Infantry* 61 (September–October 1971): 6–11.

Ney, Virgil. *The Evolution of the Armored Infantry Rifle Squad.* Fort Monroe, VA: Headquarters, U.S. Army Combat Developments Command, 1965.

Nihart, Brooke. "Army Moves Out on New Armored Vehicles, Gun." *Armed Forces Journal International* 108 (3 May 1971): 34–35.

Nihart, Brooke. "FMC Gets MICV Development." *Armed Forces Journal International* 110 (June 1973): 18.

Nobles, David L. "Light Armored Cavalry—The Right Force at the Right Time." *Armor* 104 (January–February 1995): 15–18.

Nobles, David L. "The Light Armored Force: An Urgent Need, A Ready Solution." *Armor* 99 (May–June 1990): 8–13.

Nordeen, Lon O., Jr. "Combined Arms Performer Endures Its Purgatory Alone." *Army* 37 (July 1987): 63.

Nordeen, Lon O., Jr."Today's Infantryman." *National Defense* 70 (January 1986): 35–40.

Nordeen, Lon. O., Jr. "The U.S. Army's Light Infantry Division: A New Rapidly Deployable Peace-Keeping Force." *Military Technology* 8 (October 1984): 53–61.

Norman, Kurt D., and George M. Schwartz. "Scout MOUT: Model for Future Cavalry Training." *Armor* 83 (September–October 1993): 36–41.

Norris, George T. "The Soviet BTR-80." *Infantry* 76 (November–December 1986): 23–24.

Northumberland, Pierre de Percin. "Is Infantry Drill Necessary?" *NATO's Sixteen Nations* 31 (June 1986): 50–52.

Notley, C. R. S. "Armour/Anti-armour: The Future." *RUSI Journal* 132 (March 1987): 17–22.

O'Ballance, Edgar. *No Victor, No Vanquished: The Yom Kippur War.* San Rafael, CA: Presidio, 1978.

O'Connell, Robert L. "Putting Weapons in Perspective." *Armed Forces and Society* 9 (spring 1983): 442.

Offley, Ron. "Close Air Support for the Air Land Battle." *Infantry* 75 (September–October 1985): 21–25.

Ogorkiewicz, R. M. "Aluminum Armour for Fighting Vehicles." *International Defense Review* 3 (June 1970): 200–202.

Ogorkiewicz, R. M. "Armored Vehicles for the Infantry." *International Defense Review* 19 (1986): 8(Suppl): 31–34.

Ogorkiewicz, R. M. "The Future of the Battle Tank." *RUSI Journal* 116 (June 1971): 22–29.

Ogorkiewicz, R. M. "MCV 80—The New British Infantry Combat Vehicle." *International Defense Review* 15 (1982): 721–26.

Ogorkiewicz, R. M. "Mechanized Infantry." *Military Review* 54 (August 1974): 67–73.

Ogorkiewicz, R. M. "A New Version of a Well-Proven APC—The Product Improved M113A1." *International Defense Review* (April 1971): 163–65.

Ogorkiewicz, R. M. "Tanks in Tomorrow's Armies." *Military Review* 54 (February 1974): 20–26.

Ogorkiewicz, R. M. "U.S. Advances in Tracked Armoured Vehicle Transmission." *International Defense Review* 11 (1978): 396–400.

Ogorkiewicz, R. M. "The XM723 Mechanized Infantry Combat Vehicle." *International Defense Review* 8 (October 1975): 695–98.

Ogorkiewicz, Richard M. *Armour: The Development of Mechanized Forces and Their Equipment*. London: Stevens and Sons, 1960.

Ogorkiewicz, Richard M. *Design and Development of Fighting Vehicles*. London: Macdonald, 1968.

Orgorkiewicz, Richard M. "FMC's Armoured Infantry Fighting Vehicle." *International Defense Review* 9 (1980): 1391–95.

Ogorkiewicz, Richard M. "French Armor." *Armor* 75 (September–October 1965): 39–44.

Ogorkiewicz, Richard M. "Future Infantry Armored Vehicles." *Armor* 87 (November–December 1978): 24–28.

Ogorkiewicz, Richard M. "Infantry Armored Vehicles: Origins and Future Trends." *International Defense Review* 23 (May 1990): 507–12.

Ogorkiewicz, Richard M. "Infantry Combat Vehicles." *Infantry* 61 (January–February 1971): 46–49.

Ogorkiewicz, Richard M. "Infantry Combat Vehicles: New Equipment for the Armored Infantry." *Armor* 70 (September–October 1961): 18–21.

Ogorkiewicz, Richard M. "Infantry's Combat Vehicles." *Armor* 83 (September–October 1974): 16–20.

Ogorkiewicz, Richard M. "Israel's Merkava Mark 2 Battle Tank." *Armor* 94 (November–December 1985): 39–43.

Ogorkiewicz, Richard M. "Naval Concepts in Armored Warfare." *Armor* 69 (May–June 1960): 22–23.

Ogorkiewicz, Richard M. *Technology of Tanks*. Coulsdon, Surrey, UK: Jane's Information Group, 1991.

Ogorkiewicz, Richard M. "The U.S. Armoured Combat Vehicle Technology Program—A Closer Look." *International Defense Review* (1979): 811–15.

Oliver, George F., III. "Light Cav Has Strategic Value Today" (letter). *Military Review* 74 (November 1993): 65.

Oliver, James E. "Using Bradley Gunnery Techniques." Armor 95 (July–August 1986): 9–10.

O'Neal, Robert J. "IRUS Recommends New Rifle Platoon/Squad for 1970–75." *Infantry* 62 (January–February 1972): 22–26.

O'Neill, R. P. H. "The Case for a Single Combat Arm." *RUSI Journal* 122 (June 1977): 35–39.

O'Neill, Timothy R. "Needed Now: An Antiarmor Doctrine." *Armor* 80 (January–February 1976): 18–24.

Osteen, John L. "Dynamic Traning for Rifle Squads." *Infantry* 61 (November–December 1971): 32–33.

Panton, Jefferson R. "Company Team Offensive Operations: An Urban Terrain." *Armor* (November–December 1993): 21–25.

Pardew, James W., Jr. "The Iraqi Army's Defeat in Kuwait." *Parameters* 21 (winter 1991): 17.

Patton, Garry L. "Mechanized Infantry—Modern Day Dragoons." *Armor* 75 (July–August 1966): 30–31.

Patton, George S., III. "Some Thoughts on Cavalry." *Armor* 88 (May–June 1979): 26–29.

Pearce, William T. "The Close Combat Armoured Vehicle." *Armor* 93 (September–October 1984): 44–45.

Pearl, Barton J. "Re-Arm M113." *Infantry* 76 (May–June 1986): 6.

Pearson, David. "Ground Mobility, Past, Present & Future." *Infantry* 62 (March–April

1972): 20–27.

Perkins, Mike S., and Craig S. Wilkinson. *Accuracy of Boresight Equipment for the 25-Millimeter Gun of the Bradley Fighting Vehicle*. Fort Benning, GA: U.S. Army Research Institute, 1988.

Perkins, Stephen P. "Standardize Combat Load." *Infantry* 76 (January–February 1986): 16–18.

Perry, Mark J., and Marc A. Sierra. "Team Eagle." *Infantry* 79 (November–December 1989): 11–13.

Pigg, Jim. "Why Cav Changed in the 70's" (letter). *Armor* (January–February 1995): 3.

Po, Enrico. "Wheeled APC's: Technologies, Market, Employment." *Military Technology* 24 (June 1981): 72–88.

Premo, Gregory J. "The BIFV and Communications." *Infantry* 73 (July–August 1983): 6–7.

Prillaman, John P. "What's Happened to the Mechanized Infantry?" *Armor* 79 (1970): 29–31.

Puryear, A. A., and Gerald R. Haywood, II. "Ar Rumaylah Airfield Succumbs to Hasty Attack." *Armor* 100 (September–October 1991): 16–20.

Radcliffe, Robert W., and Greg S. Burnett. "Bradley Gunnery." *Infantry* 76 (May–June 1986): 35–37.

Rasor, Dina. *The Pentagon Underground*. New York: Times Books, 1985.

Ray, Arjun. "Experiences of a BMP Battalion Commander—Reform and the Way Ahead for Armoured Infantry." *RUSI Journal* 134 (1989): 32–38.

Ray, Max D., and Robert L. Schmidt. "Mechanized Infantry Formations—Another Look." *Infantry* 63 (November–December 1973): 33–35.

Record, Jeffrey. "Armored Advance Rates: A Historical Inquiry." *Military Review* 53 (September 1973): 63–72.

Rehrer, Raymond. "Merkava Debate" (letter). *Armor* 95 (May–June 1986): 2.

Reid, Brian Holden. "Major General J. F. C. Fuller and the Problem of Military Movement." *Armor* 100 (July–August 1991): 26–31.

Reiss, David W. "Bradley Training Devices." *Infantry* 76 (May–June 1986): 38–39.

Rhoderick-Jones, Robin. "A Soldier's Tank." *RUSI Journal* 120 (September 1975): 30–34.

Rich, Marvin. "Source Selection Scheme for Scout System." *Defense Industry Bulletin* 7 (winter 1971): 19–21.

Riggsby, R. M. "Armored Infantry Battalion Training Tests in the 3rd Armored Division." *Armor* 66 (July–August 1957): 17–21.

Riley, J. P. "MCV-80 and Beyond—Implications for the Infantry." *RUSI Journal* 13 (September 1986): 23–28.

Ritz, Alfred. "Panzergrenadiere: German Mechanized Infantry." *Armor* 82 (July–August 1973): 20–24.

Robel, Michael K. "Bradley Battalions Need Tanks, Not ITVs." *Armor* 95 (January–February 1986): 48–49.

Robel, Michael K. (Untitled letter in re Flynn and Miller). *Armor* (November–December 1993): 2.

Robel, Michael K. "Operational Mobility for the Light Division." *Military Review* 69 (July 1989): 41–48.

Roberts, William L. "Employment of Armored Infantry of an Armored Division." *Military Review* 23 (May 1943): 33–37.

Robertson, Jarrett J. "Cavalry Missions and Structure." *Armor* 97 (November–December 1988): 30–33.

Robinett, Paul McDonald. *Armor Command.* Washington, DC: McGregor and Werner, 1958.

Rogge, R. E. "A Jaundiced View of Tanks." *Armor* 91 (September–October 1982): 15–17.

Rogge, Robert E. "Over-Burdened" (letter). *Infantry* 77 (May–June 1987): 5.

Romjue, John L. *From Active Defense to AirLand Battle: The Development of Army Doctrine, 1973–1982.* Fort Monroe, VA: Historical Office, United States Army Training and Doctrine Command, 1984.

Romjue, John L. *A History of Army 86.* Fort Monroe, VA: Historical Office, United States Army Training and Doctrine Command, 1983.

Rosen, Stephen Peter. *Winning the Next War: Innovation and the Modern Military.* Ithaca, NY: Cornell, 1991.

Rosenau, William. *Live Fire: Testing the U.S. Army's Bradley Fighting Vehicle, Case Program.* Cambridge, MA: Harvard University: John F. Kennedy School of Government, 1988.

Rosenberger, John D., and Thomas E. White. "The Armored Cavalry Regiment: Catalyst for Operational Success." *Armor* 95 (March–April 1986): 11–15.

Rosenberger, John D., and Thomas E. White. "Cavalry in the AirLand Battle." *Armor* 95 (January–February 1986): 22–23.

Rossetto, Luigi. "Armored Infantry?" *Army Quarterly* 109 (April 1979): 147–52.

Rouse, Edward N. "Platoon Teams." *Infantry* 76 (July–August 1986): 4.

Rozman, Thomas R. "The Mechanized Rifle Company as a Leadership Academy." *Infantry* 81 (July–August 1991): 31.

Rozman, Thomas R. "Thoughts on Medium or Motorized Forces." *Infantry* 81 (January–February 1991): 22–26.

Rozman, Thomas R., and Edward E. Blankenhagen. "An Infantry Mount for the 21st Century." *Infantry* 80 (September–October 1990): 21–24.

Rozman, Tom. "Making Light Forces More Flexible and Responsive." *Armor* 100 (January–February 1991): 18–20.

Rutherford, Andrew. "Combat Assault Vehicle." *Armor* 89 (November–December 1980): 49.

Ryan, Michael P. "Dismounted Training Day." *Infantry* 78 (March–April 1988): 28–29.

Ryazantsev, D. "Infantry Fighting Vehicle." *Soviet Military Review* 11 (November 1979): 24–26.

Rybicki, John. "A Formula for Defence." *Military Technology* 8 (March 1984): 92–103.

Salerno, George. "Repairing the Broken Sabre: Overview of L-Series Divisional Cavalry." *Armor* (January–February 1994): 29–34.

Scales, Robert H., Jr. *Certain Victory: United States Army in the Gulf War.* Washington, DC: Office of the Chief of Staff, United States Army, 1993.

Schad, Dave. "The Bradley Short Course." *Soldiers* 40 (November 1985): 20–22.

Schneider, Christopher F. "Define the Scout's Mission." *Armor* 89 (March–April 1980): 2.

Schneider, Christopher F. "Thoughts on Heavy vs. Light" (letter). *Armor* 96 (July–August 1987): 2–3.

Schneider, Wolfgang. "Marder A3—Upgrades of a Proven ICV." *International Defense Review* 22 (November 1989): 76–88.

Scholtes, Richard. "Where Have All the Infantrymen Gone?" *Armed Forces Journal International* 124 (October 1986): 92–97.

Schweiger, Michael J. "Practical POL." *Infantry* 73 (January–February 1983): 24–27.

Scribner, Barry. "HMMWVs and Scouts: Do They Mix?" *Armor* 98 (July–August 1989): 33–38.

Sedar, Robert P. "Employing the IFV." *Infantry* 71 (September–October 1981): 33–37.

Seehaus, Gunther. "Fire Protection in Armoured Vehicles." *Military Technology* 11 (August 1987): 50–59.

Sendak, Theodore T., and Kevin B. Smith. "History, Synergy, and the Cavalry Squadron." *Military Review* 71 (November 1991): 57–66.

Senger und Etterlin, F. M. von. "The New German Armored Carrier." *Armor* 79 (January–February 1970): 18–21.

Senger und Etterlin, F. M. von. *Tanks of the World, 1983*. Translated by Richard Simpkin. Annapolis, MD: Nautical + Aviation Publishing Company of America, 1983.

Shackleton, N. A. "The Motorized Infantry Threat." *Canadian Army Journal* 17 (June 1963): 16–27.

Shepherd, Richard D. "Antitank Defense: A Centralized Approach." *Infantry* 63 (March–April 1973): 21–24.

Sheridan, Stan R. "MICV Now TBAT-II" (letter). *Armed Forces Journal International* 115 (June 1977): 4.

Sheridan, Stan R. "MICV TBAT II." *Armor* 86 (July–August 1977): 35–38.

Shupe, Martin L. "M1A1, Bradley Defied the Critics." *Armor* 100 (May–June 1991): 12.

Siegel, Kenneth A. "We Need Real Combined Arms Training." *Armor* 92 (January–February 1983): 46–47.

Siegerst, Robert H. "Infantry of the 70's." *Infantry* 63 (March–April 1973): 12–17.

Siegmann, Ken. "Bradley Tank to Undergo Scrutiny." *Peninsula Times Tribune* (Palo Alto, CA), 18 October 1986.

Silverman, G. A. "Dragon Maintenance Management." *Infantry* 76 (July–August 1986): 34–37.

Simchenkov, P. [IFV's.] *Soviet Military Review* 5 (May 1977): 11–17.

Simpkin, Richard [E.]. *Antitank: An Airmechanized Response to Armored Threats in the 90s*. Oxford: Brassey's, 1982.

Simpkin, Richard E. "The Broad and Narrow of It." *Armor* 92 (July–August 1983): 23–25.

Simpkin, Richard E. "Command from the Bottom." *Infantry* 75 (March–April 1985): 34–37.

Simpkin, Richard [E.]. "Countering the OMG." *Military Technology* 8 (March 1984): 82–92.

Simpkin, Richard [E.]. "FLYING TANKS?—A Tactical-Technical Analysis of the Main Battle Air Vehicle Concept." *Military Technology* 8 (August 1984): 62–80.

Simpkin, Richard [E.]. "From Array to Disarray? Tactical Aspects of Active and Reactive Armours." *Military Technology* 10 (April 1986): 18–28.

Simpkin, Richard [E.]. "The Future Anti-Armour Threat—And Possible Counters." *Military Technology* 10 (January 1986): 129–37.

Simpkin, Richard E. "The Future of Swedish Armor." *Armor* 91 (July–August 1982): 11–16.

Simpkin, Richard [E.]. "Guns for Light AFV's." *Military Technology* 10 (July 1986): 18–25.

Simpkin, Richard E. "The Heavy Force/Light Force Mix-Up." *Armor* 94 (July–August 1985): 37–41.

Simpkin, Richard [E.]. "The Infantry Fighting Vehicle: Maid-of-All-Work or Crown Princess?" *Military Technology* 9 (October 1985): 55–65.

Simpkin, Richard [E.]. "Light Armored Vehicle Armaments." *Military Technology* 9 (March 1985): 54–58.

Simpkin, Richard [E.]. "The Light Tank: A Viable Proposal?" *Military Technology* 6 (August 1982): 92–108.

Simpkin, Richard E. "Living with Tanks." *Armor* 93 (March–April 1984): 20–23.

Simpkin, Richard E. *Mechanized Infantry*. Oxford: Brassey's, 1980.

Simpkin, Richard [E.]. "Multi-layer Armour: A Quantum Jump." *NATO's Fifteen Nations* Special Issue No. 1 (1981): 54–56.

Simpkin, Richard E. "A New Proposal for Fighting Vehicles." *Armor* 89 (November–December 1980): 13–17.

Simpkin, Richard [E.]. *Race to the Swift: Thoughts on Twenty-First Century Warfare*. London: Brassey's, 1985.

Simpkin, Richard [E.]. *Red Armour: An Examination of the Soviet Mobile Force Concept*. London: Brassey's, 1984.

Simpkin, Richard [E.]. "The Renaissance of the Light Combat Vehicle." *NATO's Sixteen Nations* 28 (June–July 1983): 89–94.

Simpkin, Richard E. "Room at the Top." *Armor* 94 (January–February 1985): 18–23.

Simpkin, Richard [E.]. "The Tank at the Fiveways." *Military Technology* 9 (May 1985): 54–65.

Simpkin, Richard E. "Tank Hunting." *Infantry* 74 (July–August 1984): 23–27.

Simpkin, Richard [E.]. "Tank or Tank Destroyer?" *Military Technology* 7 (May 1983): 14–33.

Simpkin, Richard [E.]. *Tank Warfare: An Analysis of Soviet and NATO Tank Philosophy*. London: Brassey's, 1979.

Simpkin, Richard E. "Task Configuration for Fighting Vehicles." *Armor* 92 (November–December 1983): 13–18.

Simpkin, Richard [E.]. "A Triumvirate of Heavyweights? One Possible Solution for the Heavy Manoevre Force." *Military Technology* 10 (October 1986): 149–58.

Simpkin, Richard E. "When the Squad Dismounts." *Infantry* 73 (November–December 1983): 14–18.

Skiles, Dale. "The Scout Platoon Revisited." *Armor* 93 (January–February 1989): 44–45.

Skull, Kenneth C. "Heavy and Light Force Integration at the National Training Center." *Military Review* 73 (May 1993): 41–52.

Smart, Donald L. "A Reply from DCD's Director to Sgt. Sundlof's Proposals." *Armor* 97 (January–February 1988): 49–50.

Smith, Edward C. "The Infantry Need for Mortars." *Military Review* 61 (December 1981): 29–31.

Smith, Robert C. "Positional Warfare?" *Armor* 90 (January–February 1981): 58–59.

Smith, Roger. "What Is a Tank?" *Armor* 93 (September–October 1984): 45–46.

Smith, Seward. *Human Factors Implications of MICV OT II for Infantry Fighting Vehicle (IFV) Development*. Alexandria, VA: Department of the Army, Office of the Deputy Chief of Staff for Personnel, US Army Research Institute for the Social and Behavioral Sciences, 1968.

Smith, Thomas T. "Blitzkrieg: The Myth of Blitz." *Infantry* 79 (July–August 1990): 28–

30.

Smith, Thomas T. "Bradley Gunnery Evolution." *Infantry* 78 (May–June 1988): 20–26.

Smith, Thomas T. "Bradley Platoon Tips." *Infantry* 76 (January–February 1986): 12–14.

Smithers, A. J. *Cambrai: The First Great Tank Battle, 1917*. London: Leo Cooper, 1992.

Smithers, A. J. *Rude Mechanicals: An Account of Tank Maturity during the Second World War*. London: Leo Cooper, 1987.

"Soviet Motorized Infantry." *Infantry* 78 (July–August 1989): 42–43.

Sparks, Mike. "M113's Maximize Mechanized Infantry Mobility and Firepower in Contingency Ops." *Armor* 104 (January–February 1995): 6–14.

Spigelmire, Michael F. "Bradley Platoon Organization." *Infantry* 80 (January–February 1990): 1.

Spigelmire, Michael F. "Infantry Combined Arms Doctrine Across the AirLand Battlefield." *Engineer* 19 (July 1989): 13–14.

Spigelmire, Michael F. "Mortars and the Combined Arms Team." *Infantry* 80 (May–June 1990): 1–2.

Spinney, Franklin C. *Defense Facts of Life: The Plans/Reality Mismatc*h. Ed. James Clay Thompson. Boulder, CO: Westview, 1985.

Spurgeon, Harold L., and Stanley C. Crist. "Armor in the 21st Century." *Armor* 103 (January–February 1994): 12–14.

Spurgeon, Harold L., and Stanley C. Crist. "The Tank Is Dead, Long Live the Tank." *Armor* 96 (March–April 1987): 45–46.

Stanton, Martin N. "Double Your Dragon." *Infantry* 79 (May–June 1989): 40–42.

Stanton, Martin N. "Scout Platoon Vehicle." *Infantry* 79 (January–February 1989): 43–44.

Stanton, Shelby L. *The Rise and Fall of an American Army: U.S. Ground Forces in Vietnam, 1965–1973*. New York: Dell, 1985.

Starry, Donn A. *Mounted Combat in Vietnam*. Washington, DC: Department of the Army, 1978.

Steele, Dennis. "Like a Stick in a Hornet's Nest." *Army* 38 (November 1988): 26–29.

Steele, Dennis. "Like a Stick in a Hornet's Nest." *Army* 38 (December 1988): 28–32.

Steele, Dennis. "Like a Stick in a Hornet's Nest." *Army* 39 (January 1989): 33–35.

Steele, Dennis. "Taking on the Hits." *Soldiers* 41 (August 1986): 10–12.

Steinhauser, Thomas C. "Manytanks." *Armed Forces Journal International* 111 (August 1974): 30–31.

Stepaniak, Frederick. "Defeating Armor." *Infantry* 63 (September–October 1973): 32–35.

Stephens, John M. "The Amazing Scouts." *Armor* 99 (May–June 1990): 7.

Sterrett, John, and William L. Shackelford. "Mounted Ground Combat Operations." *Armor* 88 (July–August 1979): 33–36.

Stevenson, James P. *The Pentagon Paradox: The Development of the F-18 Hornet*. Annapolis: Naval Institute Press, 1993.

Stewart, Duncan F. "The Balanced Combined Arms Battalion." *Armor* 84 (July–August 1975): 35–40.

Stubbing, Richard A. *The Defense Game: An Insider Explores the Astonishing Realities of America's Defense Establishment*. New York: Harper and Row, 1986.

Stubbs, Kevin D. "Beyond the Army of Excellence." *Military Review* 68 (August 1988): 24–41.

Sullivan, R. F. "Soviet APC's Reveal a Shift in Tactics." *Infantry* 61 (November–

December 1971): 30–31.

Sullivan, Roy F. "The APC in RVN." *Armor* 80 (1970): 36–38.

Sullivan, Roy F. "Muddling Along with the M113." *Armor* 81 (May–June 1971): 9–12.

Sundlof, Russ. "Modifying the Army's Armored Vehicles." *Armor* 97 (January–February 1988): 48–49.

Sutherland, John R., III. "Innovations from Operation DESERT STORM." *Infantry* 85 (May–June 1995): 15–19.

Swinburn, Richard. "Future Armored Warfare: The Case for the Tank." *RUSI Journal* 137 (June 1992): 35–37.

Taber, Robert C. "MISPC: A Challenge for the Squad Leader." *Infantry* 61 (July–August 1971): 6–9.

"Tactical Utilisation of a Light Fighting Unit." *International Defense Review* (1968): 184–87.

Tal, Israel. "Thoughts on the Future of Armoured Forces." *International Defense Review* (1983): 777–79.

Talbott, Orwin C. "Gun or Guided? An Evaluation of Our Antiarmor Weapons." *Infantry* 63 (March–April 1973): 6–11.

"Tank-Automotive News: Combat Vehicle Testing." *National Defense* 59 (September–October 1979): 80.

Thiele, H. J. "*Panzergrenadiere*." *Infantry* 53 (November–December 1963): 38–43.

Thiele, H. J. "The West German Infantry." *Infantry* 52 (January–February 1962): 8–10.

Thomas, Kent D. "Employing a Brigade Scout Platoon." *Armor* 99 (November–December 1990): 45–47.

Thompson, Loren B., ed. *The Defense Beat: The Dilemmas of Defense Coverage.* New York: Lexington, 1991

Thompson, Paul E., Jr. "Light/Heavy Integration at the Joint Readiness Training Center." *Armor* 107 (July–August 1998), 10–14.

Thompson, Robert H. "Lightweight Combat Vehicles: Can They Make It on the Modern Battlefield?" *Armed Forces Journal International* 117 (June 1980): 54.

Tiffany, Allen L. " 'Light' Infantry Division with More for the Fight." *Military Review* 71 (August 1991): 40–55.

Tillman, Andrew C. "Extending the Range of IFV's—The 30-mm Bushmaster II." *International Defense Review* (July 1992): 685–87.

Trautman, Konrad. "Misusing the TOW." *Infantry* 73 (July–August 1983): 12–13.

Treat, Harlie R. "Machinegunners." *Infantry* 73 (November–December 1983): 38–39.

Truett, Michael V. "Rail-Loading a Heavy Brigade." *Infantry* 76 (November–December 1986): 16.

Trythall, Anthony John. *"Boney" Fuller: Soldier, Strategist, and Writer, 1878–1966.* New Brunswick, NJ: Rutgers University Press, 1977.

Tsouras, Peter G. *Changing Orders: The Evolution of the World Armies, 1945 to the Present.* New York: Facts on File, 1994.

Turk, Paul. "The Hughes 'Chain Gun' Concept." *Military Technology* 6 (May 1982): 37–42.

Tuttle, George A. "Fighting Vehicles." *Armor* 88 (January–February 1979): 43.

Uhle-Wettler, Franz. "Infantry versus Tank." *NATO's Sixteen Nations* 29 (June–July 1984): 49–53.

Undergriff, Donald E. "Scout Platoon Offensive Reconnaissance." *Infantry* 78 (September–October 1988): 15–17.

United States. Army. Chief of Staff. *Bradley Fighting Vehicle: White Paper 1986.* Washington, DC: USGPO, 1986.

United States. Army. Command and General Staff College. *Selected Readings in the Development of Combat Divisions.* Fort Leavenworth, KS: Army Command and General Staff College, 1971.

United States. Army. Command and General Staff College. Combat Studies Institute. *Selected Readings in Military History: Evolution of Combined Arms Warfare.* Fort Leavenworth, KS: U.S. Army Command and General Staff College, 1983.

United States. Army. Infantry School. *Infantry Reference Data, ROAD.* Fort Benning, GA: USAIS, 1962.

United States. Army. Infantry School. *Reference Book: Infantry, and Airborne Battle Groups.* Fort Benning, GA: USAIS, 1961.

United States. Army. Operational Test and Evaluation Agency. *U.S. Army Operational Test and Evaluation Agency Annual Historical Review, 1 October 1978–30 September 1979.*

United States. Army. Tank-Automotive Command, Warren MI. *Annual Historical Summary, FY1970.* Compiled by J. L. Avesian, Command Historian. Warren, MI: U.S. Army Tank-Automotive Command, 1971.

United States. Army. Tank-Automotive Command. Historical Office. Warren, MI. *U.S. Army Tank-Automotive Command. Annual Report of Major Activities. FY1974.* Compiled by J. L. Avesian. Warren, MI: United States Army Tank-Automotive Command, 1975.

United States. Army. Tank-Automotive Materiel Readiness Command. Warren, MI. *U.S. Army Tank-Automotive Materiel Readiness Command Annual Historical Review. 1 October 1978 to 30 September 1979.* Compiled by Joseph L. Avesian. Warren, MI: U.S. Army Tank-Automotive Command,1980.

United States. Army. Tank-Automotive Material Readiness Command. Historical Office. Warren, MI. *Annual Historical Review (RCS CHSISC (R3)* [*sic*], *Tank-Automotive Materiel Readiness Command 1 October 1976–30 September 1977.* Compiled by Joseph R. [*sic*] Avesian. Warren, MI: United States Army Tank-Automotive Command, 1978.

United States. Army. Weapons Command. *Technical Development Plan, Mechanized Infantry Combat Vehicle, 1965.* Rock Island, IL: U.S. Army Weapons Command, 1966.

United States. Army. Weapons Command. *Technical Development Plan, Mechanized Infantry Combat Vehicle, 1970.* Rock Island, IL: U.S. Army Weapons Command, 1966.

United States. Congress. General Accounting Office. *Operation Desert Storm. Early Performance Assessments of Bradley and Abrams.* Washington, DC: USGPO, 1992.

United States. Congress. House of Representatives. Committee on Appropriations. *Department of Defense Appropriations for 1980.* Washington, DC: USGPO, 1979.

United States. Congress. House of Representatives. Committee on Armed Services. *Defense Department Authorization and Oversight Hearings on H.R. 4428: Department of Defense Authorization of Appropriations for Fiscal Year 1987 and Oversight of Previously Authorized Programs.* Washington, DC: USGPO, 1987.

United States. Congress. House of Representatives. Committee on Armed Services. *HASC Staff Inquiry into the Bradley Live Fire Test Program.* Washington, DC: HASC, 1986.

United States. Congress. House of Representatives. Committee on Armed Services. *Hearings on Military Posture and H.R. 10929. Department of Defense Authorization for Appropriations for Fiscal Year 1979. Research and Development, Title II.* Washington, DC: USGPO, 1978.

United States. Congress. House of Representatives. Committee on Armed Serrvices. *Hearings on Military Posture and H.R. 1872 [H.R. 4040] and H.R. 2575 [S.429] and H.R. 3406.* Washington, DC: USGPO, 1979.

United States. Congress. House of Representatives. Committee on Armed Services. *Hearings on Military Posture and H.R. 6495 [H.R. 6974]: Department of Defense Authorization for Appropriations for Fiscal Year 1981.* Washington, DC: USGPO, 1980.

United States. Congress. House of Representatives. Committee on Armed Services. *Hearings on Military Posture and H.R. 2970 [H.R. 3519]: Department of Defense Authorization for Appropriations for Fiscal Year 1982.* Washington, DC: USGPO, 1981.

United States. Congress. House of Representatives. Committee on Armed Services. Acquisition Policy Panel. *Department of Defense Live Fire Testing Policy.* Washington, DC: USGPO, 1988.

United States. Congress. House of Representatives. Committee on Armed Services. Procurement and Military Nuclear Systems Subcommittee. *Department of Defense Reports Required by Fiscal Year 1988 Authorization Act on Live-Fire Testing of the Bradley Fighting Vehicle.* Washington, DC: USGPO, 1988.

United States. Congress. House of Representatives. Committee on Armed Services. Procurement and Military Nuclear Systems Subcommittee. *Swim Capability of the Bradley Fighting Vehicle.* Washington, DC: USGPO, 1988.

United States. Congress. House of Representatives. Committee on Armed Services. Research and Development Subcommittee. *Department of Defense Test Procedures.* Washington, DC: USGPO, 1986.

United States. Congress. House of Representatives. Committee on Energy and Commerce. Subcommittee on Oversight and Investigations. *Capability of the Bradley Fighting Vehicle.* Washington, DC: USGPO, 1987.

United States. Congress. Senate. Committee on Armed Services. *Department of Defense Authorization for Appropriations for Fiscal Year 1980.* Washington, DC: USGPO, 1979.

United States. Congress. Senate. Committe on Armed Services. *Department of Defense Authorization for Appropriations for Fiscal Year 1981.* Washington, DC: USGPO, 1980.

United States. Congress. Senate. Committee on Armed Services. *Fiscal Year 1978 Authorization for Mililary Procurement, Research, and Development, and Active Duty, Selected Reserve, and Civilian Personnel Strengths.* Washington, DC: USGPO, 1977.

United States. Department of Defense. *Conduct of the Persian Gulf War.* Washington, DC: USGPO, 1992.

United States. Department of the Army. *Cavalry: FM 17-95 C2.* Washington, DC: Headquarters, Department of the Army, [1977], 1981.

[United States. Department of the Army.] *FM 7-7: The Mechanized Infantry Platoon and Squad (APC).* Washington, DC: USGPO, 1985.

United States. Department of the Army. *The Mechanized Infantry Platoon/Squad.* Washington, DC: Headquarters Department of the Army, 1977.

United States. Department of the Army. *Operations, FM 100-5*. Washington, DC: Headquarters, Department of the Army, 1986.

United States. Department of the Army. *The Tank and Mechanized Infantry Company Team, FM 71-1*. Washington, DC: Headquarters, Department of the Army, 1977.

United States. Department of the Army. Headquarters. *Rifle Company, Infantry, Airborne Infantry, and Mechanized Infantry: FM 7-11*. Washington, DC: Headquarters, Department of the Army, 1962.

United States. General Accounting Office. *Army Procurement: FMC's Quality Controls and Pricing Practices on the Bradley Fighting Vehicle*. Washington, DC: USGAO, 1990.

United States. General Accounting Office. *The Army's Proposed Close Combat Armored Vehicle Team*. Washington, DC: USGAO, 1988.

United States. General Accounting Office. *Bradley Vehicle: Army's Efforts to Make It More Survivable*. Washington, DC: USGAO, 1986.

United States. General Accounting Office. *Bradley Vehicle: Comparison to the M113A3 Armored Personnel Carrier*. Washington, DC: USGAO, 1987.

United States. General Accounting Office. *Bradley Vehicle: Concerns about the Army's Vulnerability Testing*. Washington, DC: USGAO, 1986.

United States. General Accounting Office. *Bradley Vehicle: Status of the Army's Survivability Enhancement Program*. Washington, DC: USGAO, 1990.

United States. General Accounting Office. *Bradley Vehicle: Status of Transmissions Reliability*. Gaithersburg, MD: USGAO, 1988.

United States. General Accounting Office. *Bradley Vehicle Test Plans: More Information Is Needed to Fully Assess Vehicle's Survivability*. Gaithersburg, MD: USGAO, 1987.

United States. General Accounting Office. *Performance of Bradley and Abrams in Operations Desert Storm*. Washington, DC: USGAO, 1992.

United States. General Accounting Office. *Potential Reductions in Tracked Combat Vehicle Programs*. Washington, DC: USGAO, 1990.

United States. General Accounting Office. Army. Budget. *Potential Reductions to M1 Tank and Bradley Fighting Vehicle Budgets*. Washington, DC: USGAO, 1986.

United States. General Accounting Office. Army. Budget. *Potential Reductions to M1 Tank and Bradley Fighting Vehicle Budgets*. Washington, DC: USGAO, 1987.

United States. War Department. *Handbook on German Military Forces*. Washington, DC: USGPO, 1945; reprinted Baton Rouge: Louisiana State University Press, 1990.

United States. War Department. *Armored Infantry Battalion: War Department Field Manual FM 17-2*. Washington, DC: USGPO, 1944.

United States. War Department. *Armored Infantry Company: War Department Field Manual FM 17-40*. Washington, DC: USGPO, 1944.

"US Army Studies FIV to Replace Bradley." *International Defense Review*, October 1998, 21.

"The U.S. Infantry: Organizational Evolution." *Infantry* 66 (July–August 1976): 10–27.

"The U.S. Infantry: The Future." *Infantry* 66 (July–August 1976): 36–39.

van Creveld, Martin. *Technology and War: From 2000 B.C. to the Present*. New York: Free Press, 1989.

Viktorov, G. M., and William F. Robinson. "Squad, Platoon, and Company Tactics—Soviet Style." *Infantry* 52 (November–December 1962): 33–35.

Vlakancic, Peter J. "The Black Snake and Pink Panther: Tank Warfare Baltic Style." *Armor* 83 (September–October 1993): 21–29.

Wagner, Robert E. "Division Cavalry: The Broken Sabre." *Armor* 98 (September–October 1989): 35.

Wagner, Robert E. "The 'V' Maneuver Technique." *Armor* 90 (March–April 1981): 14–15.

Walmer, Max. "Modern Trends in AIFV Design." *Military Technology* 7 (September 1983): 14–27.

Walter, Enno. "Armored Infantry." *NATO's Fifteen Nations* Special Issue No. 1 (1981): 18.

Ward-Harrison, J. M. D., and J. J. H. Simpson. "A Single Fighting Arm: A Case for Integrating the Royal Armoured Corps and Infantry. *RUSI Journal* 112 (August 1967): 213–56.

Warner, Gilbert F. "Improved Mortar Vehicle." *Infantry* 79 (July–August 1989): 17–19.

Wass de Czege, Huba: "More on Infantry." *Infantry* 76 (September–October 1986): 13–15.

Wass de Czege, Huba. "Three Kinds of Infantry." *Infantry* 75 (July–August 1985): 11–13.

Wass de Czege, Huba, and Michael V. Cuff. "Improving the Demand Side of Fire Support." *Military Review* 73 (November 1993): 41–53.

Webb, George S. "More Cavalry for the Infantry Division." *Military Review* 49 (January 1964): 14–21.

Webb, Harold W. "Concepts and Technology for the Soldier." *Infantry* (July–August 1994): 15.

Weeks, J. S. "Gun or Guided Weapon?" *Ordnance* 55 (May–June 1971): 525–55.

Weeks, John. *Men Against Tanks: A History of Anti-Tank Warfare*. London: David & Charles, 1975.

Weller, Jac. "Armor and Infantry in Israel." *Military Review* 57 (April 1977): 3–11.

Weller, Jac. "Armor for Infantry." *Ordnance* 55 (January–February 1971): 353–58.

Weller, Jac. "*Bundeswehr* Infantry Organization and Tactics." *Military Review* 51 (February 1971): 82–93.

Weller, Jac. "Mechanized Infantry in Israel." *Infantry* 62 (September–October 1972): 32–38.

Weller, Jac. "Mideast Infantry Weapons." *National Defense* 39 (March–April 1975): 371–76.

Wetzel, Sam. "A Report to the Infantry Community." *Infantry* 73 (July–August 1983): 5.

Whalen, Donald P. "The Bradley MICV." *Defense Update International* 40 (1983): 5–14.

Wheeler, G.P.M.C. "The Square Regiment: A Proposal for the Re-organization of the Armour and Infantry of the British Army." *Army Quarterly* 97 (October 1968): 53–64.

Wheldon, Craig B. "Light Cavalry: Strategic Force for the Future." *Military Review* 71 (April 1993): 14–20.

White, Jerry A. "Bradley Modernization: Making a Great System Better." *Infantry* 83 (November–December 1993): 1–2.

Whitehead, John B. "Cavalry in Perspective." *Armor* 95 (March–April 1986): 44–45.

"Who's Fooling Whom?" *Armed Forces Journal International* 114 (January 1977): 7.

Wickham, John A. *John A. Wickham Jr., General, United States Army, Chief of Staff, June 1983–June 1987*. [sl:sn], 1987.

Willey, Barry E. "Where's the Commander?" *Infantry* 73 (July–August 1983): 7.

Williamson, Joel E. "Command and Control." *Infantry* 76 (May–June 1986): 25–29.

Wilson, Dale. "The Infantry's Armor." *Soldiers* 34 (August 1979): 11–12.

Wilson, Dale E. "Don't Subdivide Infantry." *Infantry* 78 (March–April 1988): 3.

Wilson, J. R. "U.S. Army Experts Assess Needs for the Next Century." *International Defense Review* 22 (September 1989): 1176–81.

Wilson, Melvin. "Brigade Cavalry Troop Would Cover Void in Brigade Dept Ops?" (letter). *Armor* 103 (January–February 1994): 2–3.

Wilson, Robert. "Tanks in the Divisional Cavalry Squadron." *Armor* 101 (July–August 1992): 6–11.

Windle, Joseph C., and Harold E. Raugh, Jr. "COHORT Company Training Program." *Infantry* 74 (November–December 1984): 26.

"The Winds of Reform." *Time*, 7 March 1983, 12–30.

Wintels, J., and Kris P. Thompson. "Tanks with Infantry." *Infantry* 80 (September–October 1990): 12–16.

Witsken, Jeff, and Lee MacTaggart. "Light Cavalry in the 10th Mountain Division." *Armor* 99 (July–August 1990): 36–40.

Wolf, Duquesne A. *The Infantry Brigade in Combat*. Manhattan, KS: Sunflower, 1987.

Wood, Jack B. "Heavy-Light Connection: Brigade." *Infantry* 74 (July–August 1984): 19–22.

Wood, John S. "Future Infantry Arms." *Ordnance* 54 (July–August 1969): 74–77.

Woods, Thomas G. "MOS 1560." *Armor* 64 (July–August 1955): 27–31.

Wottlin, Robert J. "The Case for Light Cavalry." *Armor* 100 (November–December 1991): 30–33.

Woznick, John C. "The Scout Vehicle." *Armor* 103 (September–October 1994): 31–34.

Wunderlich, Tony. "Lucky Scouts Dodge 'Big Bullets' That Ripped Their Bradley." *Armor* 100 (1991): 22–23.

Wyszecki, Fritz von. "Armored Warfare Trends in Central Europe." *Military Review* 48 (November 1968): 76–79.

"XM-2 and 3." *Armor* 88 (May–June 1979): 30–31.

Yarrison, James L. "*Auftragstaktik*" (letter). *Infantry* 80 (January–February 1991): 3.

Yeargan, Robert G. "The Modern U.S. Infantry." *Infantry* 67 (July–August 1977): 11–15.

Younts, James O. "Antiarmor Tactics in a Built-Up Area." *Infantry* 63 (January–February 1973): 22–23.

Zaloga, Steven J. *The M2 Bradley Infantry Fighting Vehicle*. London: Osprey, 1986.

Zaloga, Steven J., and James W. Loop. *Modern American Armour: Combat Vehicles of the United States Army Today*. London: Arms and Armour, 1982.

Zeigler, Robert P. "Mechanized Fighting Vehicle." *Military Review* 46 (July 1966): 3–12.

Zeller, Phillip J. "Where's the Div Cav [*sic*] Squadron?" (letter). *Armor* (July–August 1988): 3.

Zick, Thomas G., Jr. "Bradley Gunnery Training." *Infantry* 76 (January–February 1986): 37–39.

Zumbro, Ralph. "Independent Operations." *Armor* 83 (September–October 1993): 15–53.

Index

About the Author

W. BLAIR HAWORTH, JR. is a contracting historian and defense analyst specializing in military history and the history of technology. He has previously participated in projects for the U.S. Congress's Office of Technology Assessment, the U.S. Department of Defense Cold War Project, and the American Maritime History Project, among others. He holds a bachelor's degree from the University of North Carolina at Chapel Hill, a master's degree in the history of science from Harvard University, and a doctorate in history from Duke University.

ISBN 0-313-30974-4
90000>
EAN
9 780313 309748
HARDCOVER BAR CODE